# THE HOP FARMER'S YEAR

### The Seasons, Tools and Methods of Hop Growers in New York State's Golden Age of Hops

Albert C. Bullard

Square Circle Press
Schenectady, New York

**The Hop Farmer's Year:**
**The Seasons, Tools and Methods of Hop Growers**
**in New York State's Golden Age of Hops**

Published by
Square Circle Press LLC
PO Box 913
Schenectady, NY 12301
www.squarecirclepress.com

© 2015 by Albert C. Bullard.
All rights reserved. No part of this publication may be reproduced or transmitted in any form or by any means, electronic or mechanical, except brief quotes extracted for the purpose of book reviews or similar articles, without permission in writing from the publisher.

First paperback edition 2015.
Printed and bound in the United States of America on acid-free, durable paper.
ISBN-13: 978-0-9856926-7-4
ISBN-10: 0-9856926-7-7
Library of Congress Control Number: 2015949931

**Publisher's Acknowledgments**
Cover ©2015 by Square Circle Press; design by Richard Vang. Black and white cover images, courtesy of the author. Color images courtesy of Richard Vang.

Unless otherwise noted, all interior images are courtesy of the author.

*This book is dedicated to all the folks who shared their family stories, their artifacts, their hop houses and their joy in remembering New York's hop heritage.*

# Contents

Acknowledgments ................................................................................................. vii

Hops & Hop Houses: An Introduction ................................................................. 3

History: The Hop in New York State ................................................................... 6

January to July: Preparation, Planting and Cultivation ..................................... 13
    Preparing the Hop Poles ............................................................................... 14
    Wire Yards ..................................................................................................... 17
    Hop Frames ................................................................................................... 21
    Setting the Hop Poles ................................................................................... 24
    Grubbing the Hops ....................................................................................... 27
    Training and Tying the Hops ....................................................................... 36
    Stringing and Winding the Hops ................................................................. 38
    Cultivating the Hops ..................................................................................... 43
    Dusting the Hops .......................................................................................... 47

August and September: Harvest and Processing ................................................ 50
    Preparing the Hop House ............................................................................. 50
    Preparing for the Pickers ............................................................................. 54
    The Harvest .................................................................................................. 58
    Hop Boxes .................................................................................................... 59
    The Hop Pickers .......................................................................................... 68
    The Box Tender ........................................................................................... 70
    Hop Knives ................................................................................................... 71
    Pole Pullers ................................................................................................... 74
    Hop Jacks ...................................................................................................... 83
    The Sacker .................................................................................................... 86
    Hop Stoves .................................................................................................... 88
    Hop Rakes, Shovels and Pushers ................................................................. 96

October to December: Baling and Selling ......................................................... 103
    Baling the Hops ........................................................................................... 104
    Screw Presses .............................................................................................. 105
    Ratchet Presses ............................................................................................ 111
    Other Presses ............................................................................................... 113

    Bale Pins and Hop Needles .................................................................................. 116
    Hop Stencils ......................................................................................................... 118
    Weighing the Hops .............................................................................................. 120
    Hop Samplers ....................................................................................................... 120
    Hop Triers ............................................................................................................ 126
    Hop Sampling Kits .............................................................................................. 128

Appendix: Hop Houses of Central New York ............................................................ 131
    Common Hop Houses ........................................................................................ 133
    Step-up Kilns ....................................................................................................... 135
    Pyramid Kilns ...................................................................................................... 137
    Round Kilns ......................................................................................................... 139
    Unique Styles ....................................................................................................... 140
    Cowls .................................................................................................................... 142
    Large Hop Farms ................................................................................................. 143

Notes on Further Reading ............................................................................................ 147

About the Author .......................................................................................................... 149

# Acknowledgments

Several people were particularly important in helping me learn so much about hop culture and history.

My father, Charles W. Bullard, who in our travels around Pennsylvania and Maryland, taught me to look for history on the land.

Dr. Bruce Buckley of the Cooperstown Graduate Program, who, in his Folklife course, first introduced me to the importance of hops in New York State, and got me away from books and out into the country.

Gary Dunbar, who shared his thesis on the status of hops in New York in the 1950s, and was always interested in my work on hops.

Bob Seaver, who gave me many hop tools, and conveyed to me his insistence on the proper use of the word "hops."

Dot Willsey, who for twenty years has been a major force in keeping hop history alive in Madison County, and has always been very encouraging of my interest in tools.

David Petri, with whom I spent many hours driving all over the hop counties of Upstate New York looking for hop houses, tools and folks that had knowledge of hop history and lore. Dave spent so much time trying to understand what folks had to say about hops and how they did the various steps through the growing year, and shared many of his ideas with me.

Mallory Arthurs, my "closer" and a student at Cooperstown Central School, who made it possible to put images with my words.

Finally, the staff at the Cooperstown High School library, who put up with my many visits.

Thank you all.

# THE HOP FARMER'S YEAR

Female and Male Hop Flowers.

# Hops & Hop Houses:
# An Introduction

The hop is one of the most enjoyable plants one may grow in their garden. They will amaze you with their vigor in the spring of the year, often growing more than a foot each day. They are rich and lush with dark green leaves as they climb upward to more than 20 feet. In late summer they are crowned with bunches of light green hops, as if they are the royalty of the garden.

As long as I have lived in Otsego County, I have enjoyed growing hops. Queenie Shultz, who lived near Hartwick Seminary, gave me my first roots. The plants made a natural sunscreen on the rear porch of our home in Westville. With a large helping of well-rotted manure, they produced large leaves and an abundance of hops in only two years.

Since that time, I have collected hop roots from several fence rows to add to my hop collection. When we later moved to our home on my wife's ancestral land, I found a hop plant in the old stone wall along one field. I marked it and very carefully moved it in the fall to the corner of our garage. The next year, I watched it grow and was anxiously anticipating the second year when the first hops would appear. The next summer I watched for the plant to flower, but noticed that no hops were being produced. Instead, a flower appeared very unlike the start of a hop, and soon it dried up and appeared to die. What went wrong?

I started to read more about the plant, and for the first time discovered that hops come in two sexes. I learned that hops have male (staminate) and female (pestillate) flowers. Since then I also learned that male vines were not very welcome in the yards. That's why they are found in the hedgerows where they were discarded many years before. Today I still have a very healthy growth of male hop vines on the corner of my garage, carefully segregated from the females that prosper in other areas in our yard.

Hops are usually set out as root cuttings. The roots have joints with small, pinkish growth nodes that sprout. These will produce a new plant if carefully planted in an upward position.

A hop has two parts to its root system. The main root has no eyes and will not propagate. It forms a growth crown and is responsible for supplying the plant with nutrients and water. The surface roots or runners grow from the crown. They were used to propagate the hop and were a source of cash to hop farmers. In a mature plant the main root is very woody and will be deep in the ground.

As the young plants start to grow, they need to be supported by poles or wires. When they reach about 20 inches in height, they start to turn as if to find some support. They climb around the pole from left to right in a clockwise direction. The hop is actually a "bine," since it has no tendrils to support it as it climbs. It simply has a twining stem that circles the pole or wire that supports it. Often twine ties hold the bines as they are trained to start their climb. Once they start upward, they grow rapidly on their own. The hops reach for the sun and will grow as far as they are able.

Since the hop is such a vigorous grower, it requires a large amount of fertilizer. The use of well-rotted manure was the favorite fertilizer in the 19th century and is still the best today.

Hops also must be kept clean of weeds, and be well cultivated as they are growing. As they grow, the surface runners must be cut back to keep the plant's strength going to the main bine.

By late July the plants will be in flower or "burr." This is the stage before the hop develops. In the weeks after the burr forms, the hop grows. Its scientific name is a "strobile." A hop has of a series of scales, called "bracts." At the base of the bracts, the yellow grains known as "lupulin" appear. The lupulin is the source of the smell and flavor that makes the hop so important to the brewing industry.

Once started, it takes about three years for the plant to reach maturity, and it will grow for many years. In the 19th century, a yard's life expectancy was usually fifteen years.

In the fall of 1966 I saw my first hop house. At that time I was a student in the Cooperstown Graduate Program, part of SUNY Oneonta. In Dr. Bruce Buckley's Folklife course, each student was given a square mile of territory in which we were to map all man-made objects. That year we were working in the Township of Springfield, Otsego County. In that township even today are many hop houses. In my square mile were two very good examples. Since I knew nothing of Central New York's great hop history, this was all new and exciting to me. Later that fall I returned to Springfield and did a detailed study of the Young hop house in my mile. From this first encounter with hop history, I have continued to learn and hunt for new information on hops.

I have always been interested in hop houses, but have since developed a special interest in the tools that were unique to hop production in the 19th and early 20th centuries. As I collected tools and related documents about them, it became evident that no one had put together a history of such tools and told how they were used. So that is my objective in this book.

This is not the usual heavily-footnoted study. I wanted to tell my story of collecting, meeting people with hop history, and to be sure that these tools are not forgotten. Generally someone told me about the use of each tool I describe, but when possible, I have patent papers and other documents that tell about the tool. Most of the tools are in my collection, or were mine at one time. In all cases I have tried to be sure that the tool was actually used on a hop farm. For those tools that are not mine, I give ownership credit on the picture. Since this is the result of only my own experiences in the hop region, others may have tools that I have never seen. That makes this a growing account of hop tools. My interest is in getting what I know and have seen written down on paper.*

From the 1960s until the early 1990s, I was primarily alone in my collecting and travel. In about 1993 a group of men came to Otsego County looking for hop tools for their museum in Washington state. They tried to buy some of my tools but I did not sell any. At about that time I teamed up with David Petri, who also had an interest in hop history, and we started to travel the hop counties in earnest. We called our effort "The Hop Project," and for about four years we were busy looking for new hop houses and new tools. All of our efforts came to a very productive conclusion in July 1995, when we presented a seminar on hops for the summer seminar program of the New York State Historical Association in Cooperstown. At that point we brought our own and many borrowed pieces of hop equipment to the week-long seminar. In the fall of that year, we also attended the first Hop Fest at the Madison County Historical Society in Oneida.

In part these efforts resulted in the acquisition of the Pope hop house at The Farmers' Museum (Cooperstown) and the exhibit, "When Hops Were King." Much of my collection was put on display for the first time. This exhibit was so popular that it lasted for five years. Since that time I have given many talks on hop history, and helped several people with the restoration of their hop houses and getting the tools needed into those kilns.

In 2001 I retired from teaching at Cooperstown Central School, and immediately started to write about my tools and experiences in the hop country. Since that time I have added a few items, but never really got back to it until 2014, and have continued to work toward the completion of this book.

---

* It should probably be stated here that unless otherwise noted in the text, the names of villages, towns and cities in this book refer to locations within New York State.

# History:
# The Hop in New York State

In New York, the hop is a native plant. As early as 1642, David Pieterszoon de Vries noted the quality of the local hops that were used in brewing by the Dutch settlers along the Hudson River. Since the first brewery in New Netherland was started in 1633, the history of the use of local hops may go back even further. These early references do not mark the start of commercial hop growing in our state, which was around 1810.

By the early 18th century, hops were being exported in small quantities from Massachusetts. Interestingly enough, we in New York were customers for some of these hops. In 1763, the schooner *Bernard* brought a cargo of 3,000 pounds of hops to New York City. By the end of the century, the area near Wilmington in Middlesex County, Massachusetts had emerged as the leading center of hop culture. By later standards, the production of 100,000 pounds in 1800 would seem very small, but it is the beginning of commercial growing. This early development of hop culture would be very important for New York, as many of the first growers in our state had New England roots.

Moving west from his home in Stow (Middlesex County, Massachusetts), James D. Coolidge started the first hop yards in New York. He settled south of the Great Western Turnpike, near Bouckville in Madison County. The commercial history of New York State hops starts here. It is very probable that others had also started hop culture to the east of Coolidge's farm, but his is the first on record.

From this beginning, hop growing rapidly spread along the Great Western Turnpike, today's Route 20. By the early 1820s Madison, Oneida and Otsego Counties were heavily involved in hops.

At that time, the distance and the difficulty of taking hops to market in Albany required a heroic effort for the frontier grower. At the market, farmers had to compete with foreign hops, which kept prices low, around $0.10 to $0.12 per pound. This was offset by the high yield per acre in the virgin soil. Some growers had yields as high as 2,000 pounds per acre.

The first boom for New York State hops was caused by a series of crop failures in England. American brewers preferred London hops in these early years. When the

English crop failed, the demand for American hops skyrocketed. As the 1820s progressed, the prices for hops increased. In 1822, a price of $1.44 per pound was recorded in Albany; this must have seemed like a fortune to the early growers. As always happens when high prices appear, everyone wanted some of the profits, and the cultivation of hops expanded. The English soon recovered and were again shipping hops. All this meant that the prices declined by the late 1820s. In 1828, the average price was only $0.10 per pound. The first attack of "hop fever" was over.

From the 1830s to the 1860s, the hop areas of Central New York grew and production soared. In 1830, about one-third of the hops grown in America were grown in New York (1,240,000 pounds). Though still behind New England, the growth of New York as a hop region was well underway. In the next ten years, New York would become the national leader, outproducing all of New England, with twice the production of Massachusetts. The dominance of New York hops was established and would remain for the rest of the century. At the same time, the center of hop culture moved to Otsego County, where it would remain for the rest of the 19th century. As the nation entered the 1860s, almost all competition was gone. New York was the source of almost 90% of the hop crop of the nation.

North entrance to the Otsego County Courthouse in Cooperstown, built in 1880. The right-hand column is decorated with carvings of hops. Left image courtesy of Richard Vang.

Many factors came together to make for this dominance. The soil and climate of New York were very good for hop culture. The newly cultivated soil yielded abundant crops. The state's location was also good for marketing the hops, and transportation systems were linking the Upstate area to the New York City market. The Erie Canal in the 1820s, steamboats on the Hudson, and the first railroads all made it easier to ship hops to market.

The linking of hop growing with the growth of the dairy industry stimulated the success of hop growing. These two agricultural activities benefited and strengthened each other. Both activities provided cash income for the farmer. Cheese and butter, not fresh milk, were continuous sources of income, while the hops would provide an extra income. The other important link between these activities was the manure provided by the cows. The hops needed large amounts of manure each season to yield well.

During the pre-Civil War years, all these factors in New York helped stimulate hop growing, but changes outside the state ensured prosperity.

American drinking habits were also changing. The consumption of beer was expanding greatly with the arrival of immigrant populations that used beer as an everyday drink. The Germans and the Irish both had beer-drinking traditions that they brought to America and greatly stimulated brewing in this nation. In 1810, only 129 commercial brewers existed in the United States. By 1850 that number had increased to 431, and by 1860 it would be 1,269 breweries. Beer was now common and much in demand wherever the new immigrants settled. The hop growers had a greatly expanded market with the proliferation of breweries.

As all these events were coming together, yet another event further helped to increase interest in hops. By the second quarter of the 19th century, a variety of agricultural publications became popular in rural America. These publications, like *The Genesee Farmer*, *The Cultivator and Country Gentleman*, and *Moore's Rural New-Yorker*, included in their pages many articles about hop culture. Lively debates were found in the letters to the editor about such questions as the best way to make a wire yard, how to fertilize the hop plants, and even the morality of growing hops. All of this information helped to improve the growing and processing of the hops, and encouraged farmers to start to grow hops, or to increase their yard size.

In the decades following the Civil War until the end of the 19th century, New York hop growing experienced its "Golden Age." During these years the yards expanded and the new technology helped the growth of hop culture. At the same time, the first signs of future decline appeared, and hop farmers saw new agricultural possibilities open.

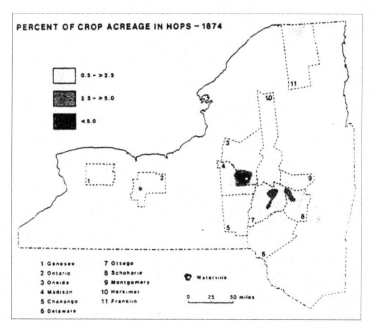

The major hop-growing counties in New York State in 1874.

In 1880 New York hop production peaked at 21,000,000 pounds, or 80% of the hop production of the nation. Most of the competition was coming from growers in California, Washington and Oregon, but New York's dominance was not seriously threatened. Prices for hops were generally good in that period, often averaging near $0.30 per pound. Since growers spent about $0.08 per pound to produce a crop, the profits were good. At times, the profits were extraordinary and a boom in growing resulted. In the winter of 1882-83, hop prices from November to February averaged over $1.00 per pound. At that price, a grower would rapidly become a wealthy man. Unfortunately, the prices did not hold at such high levels. By the 1890s, with increasing competition from the West, prices were very modest. With the onset of a national depression in 1894, prices dropped to around $0.10 per pound, or even lower. As the decade of the 1890s ended, the prices recovered somewhat, but the golden days were in the past.

During these years, throughout the hop-growing counties of Otsego, Oneida, Madison and Schoharie, the pattern of small independent growers was the rule. In my study of hop production in Milford Township in Otsego County, I found that in 1875, hop farmers averaged only 3.8 acres of hops per farm. This small production unit was the common pattern. As hops are very labor intensive the small farmer could hardly handle much more. For most farmers, hop growing was only one activity in the total farm work cycle.

An examination of the farm of Albert M. Martin in 1875 clearly illustrates the diversity of the farmer's activities. Martin had 8 acres of hops that yielded 1,200 pounds, or 150 pounds per acre. These hops would have been a good source of cash, but they were supplemented by apple production of 300 trees, 200 pounds of maple sugar, 25 gallons of maple molasses, 140 pounds of wool, 300 pounds of butter, 100 pounds of bees wax, and 23 cords of wood. This is not a unique farm, but a very normal example of agriculture in Central New York in the late 19th century. Hops would be a significant cash crop, but the farm family could survive if the prices for hops were off, or a poor crop was harvested.

After the Civil War the hop farmer, like other farmers, saw an increase in the new technology available for agriculture. Farming in the second half of the 19th century was going through a revolution in the use of machinery. In hop growing, the advent of the newly patented Harris Press is a good example of the change. This ratchet-type press increased the efficiency of baling over the earlier screw-type press. Other new tools, such as pole pullers, picking machines, improved hop boxes and better methods of setting out the yards, all helped to change the face of hop growing.

By 1900, New York had peaked as a hop-growing region. Its best days were over. In 1890, the state produced only 50% of the nation's hops. In 1900, with about 17,000,000 pounds, the percentage had dropped to 35%. It was clear that many problems were facing New York growers. In the West Coast states, the yield per acre was so much greater than in New York. Yields of 1,000 pounds per acre or higher were common in the western states, while the New York average was 630 pounds per acre.

**HOP PRODUCTION** in thousands of pounds

| | 1840 | 1850 | 1860 | 1870 | 1880 | 1899 |
|---|---|---|---|---|---|---|
| Maine | 37 | 40 | 103 | 297 | 48 | - |
| Massachusetts | 255 | 122 | 111 | 62 | 10 | 7 |
| New Hampshire | 243 | 257 | 130 | 99 | 24 | - |
| Vermont | 48 | 288 | 639 | 528 | 109 | 4 |
| NEW YORK | 447 | 2,536 | 9,672 | 17,559 | 21,629 | 17,332 |
| Wisconsin | - | 16 | 136 | 4,630 | 1,067 | 165 |
| California | - | - | - | 625 | 1,444 | 10,125 |
| Washington | - | - | - | 6 | 703 | 6,814 |
| Other | 209 | 278 | 201 | 1,651 | 1,512 | 14,763 |
| United States | 1,239 | 3,497 | 10,992 | 25,457 | 26,546 | 49,210 |

Source: U.S. Censuses of 1840 through 1900.

This table shows the dominance of New York State in hop production during the 19th century.

New York State's farmland was worn out and could not compete with the rich western soils. Transportation changes also had a great impact. The building of transcontinental railroads made it easy to ship the western hops to the expanding brewery market of the Midwest.

The growers in the East also faced some problems that were very uncommon in the West. Long before 1900, New York growers had to deal with insects and blights that were seldom seen in the West. This problem would become particularly bad in 1909, when the blue mold swept over the New York hop region. The blue mold reached its peak in 1912 when, in some areas, the hops were not harvested, but simply plowed under. Through 1913, the blight was still very common.

On top of these problems was added an infestation of hop aphids in 1914. Hops were becoming a very unattractive crop. The price of hops also remained low, in the range of $0.10 to $0.20 per pound. At this point many farmers gave up the labor-intensive crop.

The attractiveness of dairy farming also contributed to the switch away from hops. Hops had always been considered a speculative crop. Now there was the new opportunity of selling milk to the big city markets, and this was a much more secure source of income than hops. The use of railroads and the new trolley lines had opened the market for direct shipment of milk to New York City. It was clear that the time had come to move on to more economically-safe agricultural activities.

During World War I, the plight of the hop grower became even worse with the rise of enthusiasm for alcohol prohibition sweeping the nation. Hops have few uses outside of brewing, so the advent of Prohibition in 1919 pushed out many of the die-hard growers. In 1920, New York produced only 723,824 pounds of hops. The commercial hop grower was about to become an endangered species.

With the end of Prohibition in the 1930s, it was hoped that hops would come back in New York, but that never happened. Only a few farmers continued to grow hops. In Schoharie County, the Pindar family near Middleburgh grew until the early 1950s. In Otsego County, some growers in Middlefield and Schuyler Lake lasted through the 1930s. In Madison County, some die-hards like the Eisman family held on into the 1950s.

For all these growers, the chief problem was to find a buyer for their crop. As the West became dominant, the brewers simply did not want to deal with the small quantity of New York hops.

The last of commercial production in New York State was in Franklin County. The area along the St. Lawrence River had a long tradition of hop growing, going back to at least the 1850s. 100 years later, hops were still being grown in the traditional pole yards, as well as the newer wire yards. In 1953, the Trout River Hop Farm had

81 acres under cultivation. In addition, other growers—such as the Hardy Farms, Jerry Marlow, Foster Child and Earl Looker—continued to struggle on into the middle of the century. Jerry Marlow's 18 acres of hops might have been the last commercial pole yards from the Golden Age of American hop culture. The last Franklin County grower of any kind was Earl Looker, near Wippleville, who grew until 1965. After that year, the only hops grown in our state were not being cultivated on a commercial scale.

"Scene at a Hop Yard Near Cooperstown," from *Frank Leslie's Illustrated Newspaper*, 1878.

# January to July:
# Preparation, Planting and Cultivation

For centuries, agriculture had a rhythm and timing linked to the changes of the seasons. In the 19th century this was much more important than today. Hops, like other farm crops, followed such a seasonal cycle. As the year progressed, the activities and the tools used in each phase of hop culture varied.

One way to find what the yearly activities of a hop farmer were is by reading the diaries of 19th-century farmers.

In Middlefield Center, Otsego County, a family that locally had a reputation as good hop growers was the Van Patten family. I selected the year 1894 from the diaries of Leon Van Patten to use as an example. This was not a special year, but a very typical expression of the activities of a hop grower in any year.

Today the success of the Van Patten family is still visible in the hop house they used on the Middlefield Center Road. This hop house started as a simple, early-style hop house. Their prosperity in the 1880s and 1890s made it necessary to add on a step-up addition to increase the drying capacity of the kiln.

The Van Patten hop house near Middlefield Center, Otsego County.

On the inside of the kiln, on its plastered walls, are large numbers of bale weights for several decades in the late 19th century. As so many bales were pressed, it is not hard to see that during baling time, things were busy.

This building also has a very interesting exterior detail. Under the eaves are Italianate-style brackets. All together, this kiln is an interesting remnant of the prosperity of the lost hop business.

Leon was well liked and respected by the people who picked hops on his farm. In a letter from Mr. William Davidson of Troy, Leon is addressed as follows: "Leon Van Patten! The great American Hop Grower." The letter goes on to thank him for the kind treatment and good times had in the last season. Mr. Davidson obviously wanted to return in the coming picking season and is letting Leon know his wishes. For many pickers, the yearly excursion to the yards was their only vacation, and a time anxiously anticipated.

## Preparing the Hop Poles

From Leon's 1894 diary it is clear that the first months of the year were important times of preparation for the anticipated growing season. During January and February, wood had to be cut for use in the drying process, and the poles for the year had to be prepared.

"Drawed hop wood all day and put it in the hop house." This was the entry for January 11. He does not indicate how much wood he put in the hop house, but a sizable quantity would be needed for drying. The kiln was heated all night and the hop stove was not very air tight, thus causing it to consume a large quantity of wood.

Of more interest are the references to hop poles. On January 8 he cut 30 poles. Throughout February, he continues to cut, and on February 2 moved 125 poles. Leon also drew poles from Cherry Valley and his neighbors. He purchased 85 poles from Henry Rathbone in Cherry Valley for $13.75. In another trip to Cherry Valley, he purchased 65 more poles from John Gilday. In addition, in early March, he paid $0.025 each for 150 poles from Will Ryan. In his diaries, he specifies he cut or purchased 455 poles, and spent a total of $17.50.

Since he also says he cut hop poles on some days and gives no number, we may assume that the actual number is even higher. All this is important since it shows what one farmer might need in a given year. When you consider the large number of growers across the hop region, the issue of poles becomes a major concern.

# Preparation, Planting & Cultivation

Child's hop yard, c. 1940. The cut hop poles can be seen on the sled.

By the 1890s, the standard pole yard had only one pole per hill of hops. Often, for the bines to grow on, twine was strung in a tent-like pattern between the poles. This would enable the bine to spread out, and permit as much sun and air as possible to reach the plants.

A single hop pole was rather long. No set length existed, but they usually were from 18 feet, to as long as 30 feet. A pole of that size has a rather substantial diameter at the base, in the area of 5 or 6 inches. A pole, entwined with hops at harvest, was heavy and difficult to handle with all the bines wrapped around it.

In the middle of the 19th century, chestnut poles were preferred. In the early pictures, these are easy to identify because they are very crooked. In some areas near swamps, cedar poles might be used, but not in the large majority of yards. As time went on, the supply of young trees for poles started to be depleted, and other sources had to be found. Obviously, other types of trees might be used for poles, and even such trees as hemlocks, which would not last long, were made into poles.

By the 1890s, hop poles were being brought in from areas outside the hop-growing region. Many of the cedar hop poles came from the North Country and Canada. As I noted above, Leon was working with cedar poles and other woods too. Since he bought poles in Cherry Valley, these might have been brought in by rail to that village.

The pole could not be used without preparation. Two improvements had to be made. First, the bark had to be removed from the pole. This process is called "rossing." The poles would be placed in a crotched stick to hold it steady. Next, a drawshave would be used to peel the bark

A poster from the period.

Y-shaped pole supports, made by blacksmiths, about 15 in.

from the exposed side of the pole. When this was completed, the pole would be turned and the other side de-barked. Several interesting Y-shaped iron devices have been found on some of the old hop farms that might have been used in this process. I have never been specifically told that they were used in this way, but they certainly would do the job of holding a pole while the bark was removed. These tools have sharp projecting points that would hold the pole firmly. They must have once been attached to handles or a wooden support.

The second process in the preparation of the poles was to sharpen the end that would go into the ground. Normally this was done with a regular axe and a chopping block. It must have been quite a chore to sharpen several hundred hop poles. To help make this job go faster, machines were developed to sharpen the poles. In 1869, Stephen V. Barns of Triangle, in Broome County, patented such a hop pole sharpener. It was similar to a large pencil sharpener, with a moving platform to push the pole into the blade of the sharpener. It also included a small buzz saw used to trim the poles. I have never seen such a machine in the hop areas, but with the large number of poles involved, it appears an improvement over hand sharpening.

In the mid-19th century, the issue of the proper number of poles to use in each hill was a hotly debated topic. The common practice until the 1870s was to use several poles in each hill. Some advocated as many as four poles, but most went with two or three poles. In an early stereoscopic image I have seen, a hop yard with a two-pole system is well illustrated. The poles used were probably chestnut since they are so crooked.

If a yard had its hills set at $6^{1/2}$ feet apart, 889 hills would be needed for each acre. In a two- or three-pole setup, this would produce the need for a large number of poles. A pole might last for several years, but still, replacement was always an issue. Making the poles extra long so that the end might be sharpened again was a good idea. Windstorms during the growing season were always a threat to knock the poles down, and possibly break them. Others might rot each year and need new points.

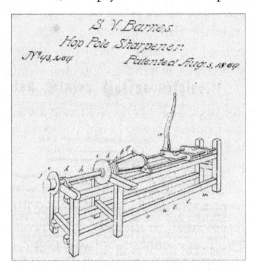

Patent drawing for a Barnes pole sharpener.

# Wire Yards

Some growers did not use poles, but preferred a wire yard. The wire yard started to appear about the middle of the 19th century. In *Moore's Rural New-Yorker* of April 1859, an article entitled, "Hop-poles - A Substitute" appears. Written by Thomas Aysworth of Herkimer County, it describes one wire system. Between posts 6 or 8 rods apart, a wire is drawn from which strings are extended to stacks in the ground. The strings each have a spring hook to attach them to the wire. To harvest the hops, the wire is lowered between one pair of posts at a time. In this way, the hops may be picked on the spot, or the strings cut from the wire and removed to pick elsewhere. The wire is kept from coming down further along by a clamp that holds it at the next pole. In the article, this system is applauded because of the "onerously expensive" cost of poles in areas depleted of poles. Also the labor saved in setting the poles each year was attractive to the author.

In *Moore's* November issue of the same year, a second article on this new style of hop yard appears. This system was a horizontal system rather than one of vertical poles. The system was described as a "low trellis-work." Several advantages were cited for this method. First, it was low enough to inspect the hops for insects and to "cleanse" the plants if insects were found. It also provided more shade to protect the tender upper shoots. The danger of wind damage was reduced, as the trellis was low. The most important advantage was in the harvesting, since the bines were not cut. This made it possible to pick only the ripe hops while letting the plant continue to grow.

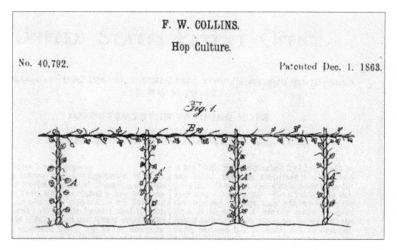

Patent drawing for the Collins low-trellis system.

Many wire yards existed, but they never achieved the popularity of the pole yard. This is reflected in the diary of Leon Van Patten in two entries for April. On April 17 he writes, "Boys took down wire yard and drawed off wire etc. etc." This was followed on April 18 with, "Boys drawed hop poles on where wire yard was and set them." It is clear that Leon was going back to the favored pole yard after trying the wire system. Unfortunately, he does not say why he made this change.

Few artifacts exist from the wire yards. On some farms old hop wire may still be found. This is little different from any heavy gauge wire. On my wife's family farm is a natural cut in the hillside that was always called the "draw." From my first visit to the farm in the late 1960s, I was told about the hop wire in the draw. Apparently plans had been made to make a wire yard, but the blue mold wiped the hops out so completely that the wire was never used. So, at some point it was dumped into the draw and is still there to this day. Coils of the wire are still above ground and easily seen, but hard to get at since trees have grown up among the coils.

Sometimes you see hooks for holding the wire on the large poles used in a wire yard. One of the first of these hooks I ever saw was on a house, and it was holding a clothes line.

The hooks came in two sizes, one at 3 inches in length, the other 4 inches. Along the side of the hooks, some are marked "Cherry Valley." These were made in the foundry of John Judd of that village. The present Cherry Valley Village Office is the building that was the Judd Foundry.

The hooks were nailed or screwed to a pole, and the wire would be held in place in the open slot of the hook. During harvest, the wire might be pushed out of the hook to let the wire down.

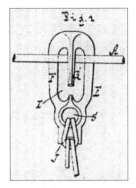

Left: Wire hooks marked "Cherry Valley" and made by the Judd Foundry. Right: Drawing for the Judd trellis hook; Patent No. 294,240, patented on Feb. 26, 1884.

*Preparation, Planting & Cultivation*

In addition, trellis hooks are sometimes found. These hooks went over the wire and were used to run lines down to the hop plants. The hook was constructed so that it could be pushed up and off the wire at harvest time to let down the bines on the strings they have ascended. One type of trellis hook was also made by John Judd and patented in 1884 (opposite page, right). If held upside down, it looks somewhat like an owl. Other cast-iron examples might be found, but the Judd hook is the only patented example I have seen.

A later type of wire hook comes from the old Moakler farm in the town of Middlefield. This hook is simply a double-bent, heavy wire hook. They used these hooks in the yards in the early 20th century and maybe as late as the 1930s.

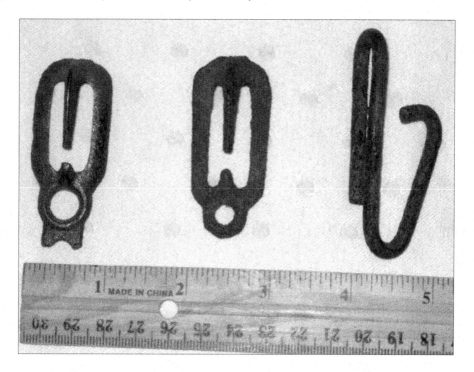

The two hooks on the left and the middle were made by the Judd Foundry. The hook on the right is a wire hook from the Moakler farm, used in the 1930s.

I have often wondered what percentage of yards would be wire and what percentage would be pole yards? An exact answer to this question will never be known, but some information is available. Many photos of hop pickers exist and in the background of these images, the type of yard may be seen. By far the largest numbers of photos show the pole yard. This would make one believe that poles were more popular.

The pole yard was less expensive and some of the poles might be cut on the home farm, as Leon did in 1894. Wire was always more costly to convert to, and farmers already had a good system, so why change?

Did the pickers prefer to work in a wire yard or a pole yard? Again I do not know and have never found anything to support either view. I do know that tradition is a hard thing to change. The pole yard was the traditional yard in New York State. When the last commercial hops were grown in the 1960s, they were grown on poles.

Did the large growers use wire yards? The size of the grower did not determine the use of a wire yard. The largest hop grower in the state at the turn of the 20th century was Jimmy Clark, with over 150 acres of hops. Clark used poles, and the photos of his yards and harvesting make this clear.

Jimmy Clark's "Hop City." The hop poles are visible along the right side of the road.

# Hop Frames

A third method of growing hops is referred to in 19th-century literature. This was the hop frame. A hop frame refers to any system that is not a simple wire yard. I have never seen any pictures of a hop frame and no artifacts of this type of system are known. The only sources we have for these systems are the patent records.

Based on the patents, two categories of hop frames were tried. The first was a system that created a frame with a combination of wood and wire supports upon which the hops could grow. Sometimes the plan was simple, as in the case of a hop frame patented in 1864 by L.S. Mason of Middlefield Center, in Otsego County. In this system, stacks with a wire running between them were placed 12 feet apart or more. In each hop hill, a stack extended 4 feet out of the ground. To this stack were attached two removable, wooden or metal training-sticks that were supported by the wire above. One advantage of this system was that it gave the hops more stability in high winds. In addition, it was easy to harvest, as all one needed to do was remove the training-sticks to get the hops down for harvest.

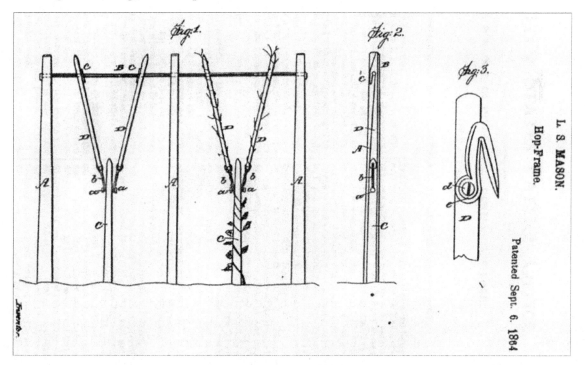

Patent drawing for a Mason hop frame.

Abram Shoemaker and Wallace Phelps of Conesville, Schoharie County, proposed a much more complex system. Their patent issued in 1868 would have totally restructured the layout of a hop yard. Their yard would be based on separate square frames with open areas in between each frame. A frame would comprise four stacks, with horizontal cross-ties securing them at the top. Attached to each post would be three pliant hoop-poles or rods that ran from the ground to the top of the support stack. This would make an arbor for the hops to grow on. With this system, twelve hop plants could grow on each frame. The advantages of such an arrangement included giving the hops full sun and good circulation of air. The separate frames would also make it difficult for insects, especially lice, to spread among the plants. If the frames were far enough apart, it would also be possible to grow corn or vegetables between the hop frames. It would seem that a system like this would be difficult to maintain and costly to put into place. In the patent application, the inventors said it had been proven as a way to stop insects, so they must have tried it.

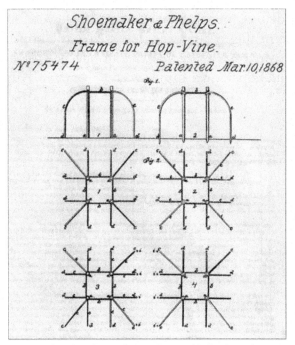

Patent drawing for a Shoemaker & Phelps hop frame.

The second category of hop frames was designed to improve on the hop pole. Several of these devices were similar to the metal clothes poles with arms that were commonly used in small yards. The idea was to have a single pole with a device to hold a series of longer poles that would extend from the main support. Clark T. Bush of Maryland, Otsego County, patented such a hop-training device in 1884. His frame was rather tall. The base pole would have been about 5 feet above ground. From it would extend four branches, each about 10 to 12 feet long. It was hoped that this method of support would let more sun and air into the hops. In addition, with this device the hop plant was cut back only to the base pole at harvesting.

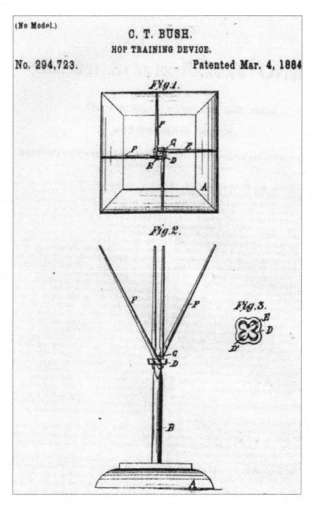

Patent drawing for a Bush hop frame.

Today little evidence exists that these inventions were ever used. In most cases, they appear to be so much more work that the savings in time and labor disappeared.

## Setting the Hop Poles

By the end of March, the frost was starting to come out of the ground and the hop yards were ready to be worked. On March 28 and 29 Leon's diary reads, "Raised hop poles all day." At the same time, the yards needed to be plowed.

All types of hop yards had to be plowed each season. This was done with a hop plow. These were rather light-duty plows, with a side-shifting beam and a clevis that made it possible to plow very close to the hop hill. It also enabled the horse to walk in the furrow or on unplowed ground. The first plowing of the season loosened and aerated the soil. It also was the start of the year's weed control.

Many factories made hop plows. One of the most celebrated was the firm of Stinge, Dexter & Coe in Munnsville, in Madison County. They made a variety of tools used in the hop industry, including plows.

Advertisement for a Munnsville hop plow.

Setting the poles might start slightly before the plowing, since the ground did not have to be as dry to set poles. The basic tool used was a hop bar. Hop bars all follow a pattern but have a variety of appearances. In an 1883 article a good description of a hop bar is given.

A convenient bar for making the hills would be about four and a half feet long, made from a rod one inch in diameter. Fifteen inches from one end it should be enlarged and formed two and a quarter inches square and gradually taper to a point, which should be steel. If the enlargement were polished, and the other end of the bar slightly enlarged it would be found to work more easily.

From the general description, the sky is the limit as far as styles of hop bars go. Looking at the collection of hop bars from the 1995 seminar at The Farmers' Museum, the variety of bars is well illustrated.

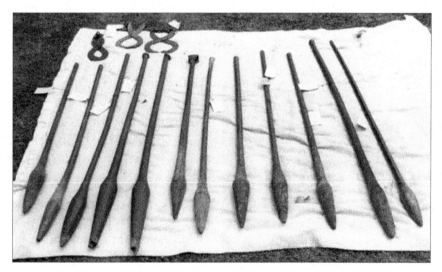

Hop bars on display at The Farmers' Museum Hop Seminar (1995).

Notice that the head, the part that goes into the ground, varies greatly in length and shape. Some bars show a difference in thickness in the bar itself. In addition, some have very distinct tops, while others have none. The bar with the very large head and long handle is a bar used in the wire yard. The support poles were larger, thus the head had to be bigger to create a larger hole. This bar is also much heavier and has a longer handle, since the hole had to be deeper.

The method of manufacture also varies among the examples. Hop bars were both cast iron and hand forged. The fourth and fifth bars from the left, from the Ripple farm in the Town of Minden in Montgomery County, were made of cast iron. Both have broken points. These were found between the studs of a hop house on that farm. It is interesting to wonder if they survived because they did break and were discarded between the studs many years ago. Hop bars are all heavy. A standard bar

Wooden-handled hop bar from Otsego County.

weight does not exist, so the bar weights vary greatly.

Another style of bar has a wooden handle and a steel point. The tip is a single conical piece welded onto the neck that holds the wooden handle. These bars are seldom seen today, and only two examples with wooden handles are known.

The bar is a very efficient tool to make a hole. The bar was driven into the ground by the force of one's hands, arms and shoulders, not by pounding it with a heavy hammer. After driving the point into the ground, it would be pushed around in the hole to enlarge the opening. This is the reason for the large head. The process would be repeated several times until the desired depth and width of the hole was created.

In the earlier years of hop culture, when several poles were in each hill, the holes had to be on an angle away from the center of the hill. In this way, the tops of the poles would be far enough apart to keep the bines from growing together at the top of the poles.

Once a hole about 2 to 3 feet deep was ready, the pole would be set into the hole. The pole would be forcefully driven into the hole by hand. It might take several tries to get a good solid setting. I have been told that the sound the pole makes going into the hole is the best way to tell if it is securely in place. Sometimes a pole did not go in well and a driving block might be used. This was a triangular wedge nailed to the pole. This could be easily struck with a heavy hammer, driving the pole into the hole securely.

The last step in setting the pole was to pack dirt tightly around the pole, and then pack it down by foot. This whole process was a very big job. All the steps were done by hand and all had to be repeated each year.

Among the hop bars pictured is another type of bar that had a separate function. In the photo of the bars on the previous page, the sixth bar from the left side is not a hop bar. This is a hop dibble, used to punch a hole in the ground to plant a hop root. The top of the bar has an open collar perpendicular to the shank of the bar. A wooden handle was placed into this collar. Using two hands, the bar was forcefully driven into the ground. Since a root only had to be several inches in, a single drive would usually make an adequate hole. This tool came from the Neff farm on Route 20, east of the Village of Madison. On the shank of the bar is an "N," indicating the owner.

The hop dibble might be used to replace a hop plant that had died in an established yard, or in the planting of a new yard.

When a new yard was created, the ground had to be prepared much as any field would be. After plowing the ground and dragging it, the grid pattern of the yard had to be laid out. Since hops grow in hills, the distance between each hill had to be marked off. Usually the hills were 7 feet apart, but the distance could vary. A simple marking device might be used, like that used to mark off rows for planting, or a rope with knots at the desired distance might be pulled tight across the field. Both methods would produce a grid pattern. Next the dibble would be used to punch a hole, and a piece of hop root or two placed in each hole. Care would be taken not to plant the root too deep. Usually about 4 inches was plenty. The growth points of the root had to point upward or the root would die. Once a yard was started, it would take three years for it to reach a productive maturity.

With an early start, the farmer might have all his poles set by the middle of April. Factors such as how fast the ground lost its frost or the amount of rain or snow could influence the time it took to complete the job. In Leon's 1894 diary, the last poles were set on April 18. After this, the next process could begin.

## Grubbing the Hops

Grubbing the hops was one of the most important tasks in the hop farmer's year. Albert W. Morse, of Eaton in Madison County, gives a general description of grubbing.

> Plow the land each way, as near the hill as can be done without breaking the roots. Remove the manure from one side of the hill; then with an implement called a "grub-hook' loosen and remove earth from around the hill to a depth of three or four inches, and pull up the surface runners, and cut them off near the hill. Also cut off the crown or top of the hill from one or two inches … Break the manure and mix with earth around the hill.

Grubbing was important for several reasons. First, it helped to control weeds in the yard. For best results, the yards had to be kept weed free. Here, the plowing and the opening of the hill would slow the early growth of weeds. Second, the cutting of the runners was necessary to keep the strength of the plant in the crown, and to keep the plant from getting out of control. Third, the mixing of the old manure from the fall was a good way to help fertilize the plant. In addition, by breaking open the hill it made it easier for the plant to start its spring growth.

The one process that is not mentioned in the above description is the removal of grubs, which gives us the term "grubbing." As the hills were opened, the farmer looked for hop grubs that could then be destroyed.

Two varieties of grubs were particularly targeted. One grub made a beetle that had "a dark hard head, and white body, with legs all on the fore part of the body. It is always found doubled up like a horse-shoe." The other was a caterpillar that would turn into a butterfly.

To help with the control of these grubs, nature sometimes lent a hand. Skunks were considered very beneficial in the hop yards, as they greatly enjoyed eating the grubs. Therefore, on the hop farm the skunk was a protected species.

Grubbing started once the land was dry enough to work. Leon started to "grub hops" on April 20, and was working at least until May 8, "Still grubbing the hops."

The tools used in this process were very basic hand tools. Grub hooks (or "hop hooks," "grubs" or "grubbers"), a knife and a hop hoe would be all the tools required. The grub hook came in a great variety of sizes and shapes. Some were manufactured, others were made at home or by the local blacksmith.

The most famous and distinctive grub was the "Otsego" grub hook. These hooks are pictured in the old literature, but no explanation of the name is given. I would assume that the style developed in Otsego County, and the name derived from this. The tool is very distinctive because of the two tines that fork off from a shaft attached to a long wooden handle. The tines arch slightly, and gracefully curve to a length of 4 to 6 inches or more. The size and the dimensions of the grubs vary, but they all have the distinctive two-tine look. Otsego grub hooks were hand-forged or factory-made. To date no example has been found with a maker's mark, but such a tool might exist.

Another style of grubber was the "horseshoe" grub hook. It also had two tines, but they were connected and remind one of a horseshoe. The tines were attached to a wooden handle, much as the Otsego grub hook was. All the horseshoe grubs I have seen were hand-forged.

Otsego grubbers.

Horseshoe grubber.

Some grubs might be considered as early examples of recycling. Many times old rasps were made over into hop grubs. An interesting example of this method is a grub with rather broad and flat tines that have a flared end. On the tines can still be seen the pattern of the old rasp that was used to make the tool.

Rasp grubber.

The foundry in Munnsville made another type of grub. In an old catalog, they refer to this tool as a "hop grub hoe." It was sold for use in the hop yards and the cultivation of grapes. This tool has a double head. On one side is a horseshoe-shaped, two-tine grub, on the other a small hoe. The tool is very heavy and has a rugged handle. Some of these Munnsvile grubs are lighter in weight, but follow the same pattern. I have never seen a marked grub of this type.

Center: Munnsville grub.

*Preparation, Planting & Cultivation*

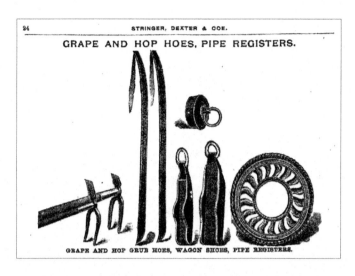

Advertisement from Stringer, Dexter & Coe of Munnsville.

Probably the most common hop grub was the simple four-tine hook. These came in two basic styles. One had an arched shank and straight flat tines, the other had tines that curved at the top. In both cases they attached to wooden handles by a pintle-shank that would be driven in to the handle, and secured by an iron collar that was hammered into place. These tools are very reminiscent of modern potato hooks.

To be sure a tool is a hop grub, one must identify it at the point of collection. The three hooks pictured below were all collected and identified as hop grubs on the Palmatier farm in the Town of Maryland in Otsego County. The larger hook is cast iron, while the two smaller hooks are hand-forged. A tool very similar to these hooks is also pictured in an article by E.O.L. of Vernon, Vermont, in Meeker's <u>Hop Culture</u> from 1883.

Four-tine grub hooks.

The "V-shaped" grub was the simplest of all the hop grubs. These were usually forged by hand out of one piece of iron. One end of the iron was shaped in to a "V," with the two tines being about 4 to 5 inches in length. The iron was then bent to an angle of slightly more than 90 degrees. This head would be driven into a wooden handle, and a simple grub would be ready to use.

V-shaped grub hook.

Sometimes very large, two-tine grubs are also found. Years ago I found such an example in Richmondville in Schoharie County. Unlike the Otsego grub, they have a very wide opening between the tines, about 4 inches. Sometimes these tools are identified as manure forks, but they have also been identified as having been used in the hop yards.

Two-tine grub.

Some farmers used single-tine hoe grubs. These were often very long, about 7 inches or more in length. The tool was flat, about 1 or 2 inches in width, with a slightly sharpened blade. I found two of these tools near Bridgewater in Oneida County, and a separate example from Springfield Township in Otsego County. Other examples of this form appear in the Upstate area, but have no hop provenance.

Single-tine hoe grubs.

When grubbing hops it was also necessary to use a knife. No special type of knife was used, but a hop knife could be used. These tools were more important in the harvesting of hops and are considered in detail in a later section.

The knife was necessary for a few reasons. First, it would be used to cut off the runners which were a spring cash crop, and then to dress the crown of the hop plant.

These runners or roots were cut into lengths, about 8 inches long, with at least two sets of eyes on each. Hop roots will keep fairly well after being dug if they are kept out of the sun and in a dry place. The eyes are very distinctive, being white with a pinkish top. When planted, it is very important to make sure that the eyes are pointing upward or the root will die. When dressing the crown of the hop, the old dead growth from last season is removed and discarded.

If a grower had an excess of roots he might sell them. On May 8 Leon recorded, "I went to Cooperstown and took about 6 bu. of hop roots to J.W. Thair." These were the roots from the grubbing that started in April. He must have had a large quantity of roots, for on May 12 he writes, "Plowed hop yard and setting out new plants." The planting of the new yard continued until June 6, when four men worked the whole day planting the new yard.

Another use of the knife in the grubbing process was to kill the grubs that might be found in the hill.

In an effort to combine the grub and the knife into one tool, a combination hop hook was patented in 1868. This tool, with either a two- or three-tine hook, had an attached knife blade on the other end of the tool. The hook would "enable the workmen to loosen the earth, bare the roots, and separate them smoothly, together with the surplus sprouts or vines, without resort to a separate tool."

Combination grubber with knife.

In the patent it also notes that "knives of various forms, attached to long handles" were in common use in the hop yards. I have seen such knives, but was always told that they were used in the harvesting of the hops, not in the grubbing process. It only stands to reason that a long-handled knife would save a lot of bending down and be much easier on one's back.

The final tool used in the grubbing was a small, three-tine fork. I have never found a good reference to these forks in the 19th-century literature, but when I collected them, the informants told me how they were used. The forks vary in size, about 4 to 8 inches in length. In the photo (opposite page, bottom) it is clear that some were hand forged, while one example was factory made. It was common to use these forks to loosen the long, fibrous hop runners from the ground. The runners would then be cut up for home use, or for sale.

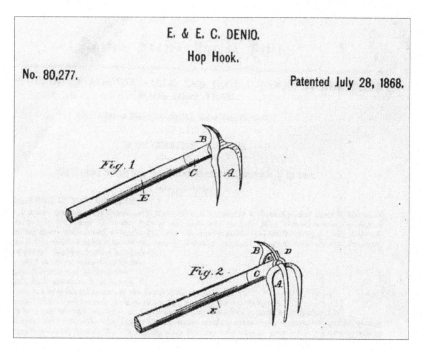

Patent drawing for a Denio hop hook.

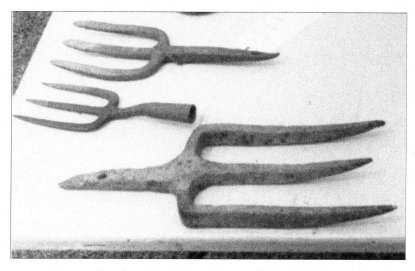

Hop forks from the collection of David Petri.

## Training and Tying the Hops

By the middle of May the hops were ready to start training up the poles and wires. Hops are very vigorous growers and produce many shoots from a single hill. When these shoots are from 18 to 24 inches high, they are ready to train up the poles or strings of a wire yard.

Considerable differences of opinion existed as to how many shoots should be trained up each pole. Some felt the number should be limited to two or three shoots at the most. This was felt to be best, since it would concentrate the vigor of the plant into those shoots. Others favored as many as six shoots. Since the hop is a vigorous grower, it should not be denied its natural growth pattern. No matter where one stood on the number of shoots, a general agreement existed that any extra shoots should be removed at the time of training. These excess shoots would be cut off and the stem covered over with earth. They would never be pulled up, as that might injure the root.

Child's hop yard in Franklin County, 1940s. The poles have just been set and some poles are still stacked from the winter.

Tying was a simple, but tedious task that required few special tools. A knife might be handy in some cases to cut the material used in tying, but no other tools were needed. Burlap was the favored material to make into string to tie the hops. A square cut piece would be hung in front of the worker. In many cases the women and children did a large share of the tying. As a new burlap string was needed it would be pulled from the piece. Also loose burlap strings might be purchased and used directly in the tying process.

It was important to get the bine started correctly up the pole or string. Hops run from right to left, or around with the sun. It is important to tie them correctly, since they will double back on themselves if started incorrectly up the pole.

The time of day for tying had be considered: "The tying should never be done on a cold day, nor in early morning as the vines are easily broken and spoiled."

If the hops were wet, it was also more difficult to tie them, as they would become scratchy and caused one's hands and arms to itch. Sometimes the bines needed a second tying, but in most cases, one tying would be all that was needed.

Wind might also interfere with the tying process. A strong wind might cause the bines to be pushed down the pole or pull the bine loose from the pole. Almost as soon as the tying was completed, the stringing of the yard would start.

A woman tying young bines at Child's hop yard, 1940s. Notice the burlap aprons used to supply the burlap strings.

# Stringing and Winding the Hops

Stringing the hop yard was a very common method and provided more growing area and better air and light for the hops in a pole yard. It was a process of running strings from one pole to the next. In the early 19th century, and well into the middle of the next, poles were cheap and abundant. Since that was the case, many poles were used in each hill. As poles became more of an expense and harder to replace, the use of strings in the hop yards became more common. Jimmy Clark of "Hop City" fame gave a good description of the stringing process.

> A popular method of stringing consists of driving a nail slightly downward into the pole only about four feet from the ground, tying the string to the top of the next pole, and so on. Drive the nail first into the first pole in the first row, then go to the second hill in the opposite row, then back to the third hill in the first row, and so on across the yard doing two rows at once. Begin by tying the twine to the first nail run the top of the twine up the next pole with a "twiner," as far as convenient carrying it around the pole and trying to catch the twine over a knot to hold it; draw up the twine close, then drop from the top of the pole down to the nail in the next pole. Step up to it and give the twine a half hitch or lop around the nail, then run the twine up to the top of the next pole, down to the next and so on across the yard … Then turn and go across in the same manner.

The most important tool used in stringing was the "stringer," "twiner," "turning pole," or "twine pole"—all different names for the same tool. Some stringers were very simple, being long basswood poles flattened on one side with a drawknife. Along this flat side, starting at the tip of the pole, several wire fence staples were nailed to run the twine through. It does not appear that a standard length was used, as some stringers were only 10 feet long, while others were as much as 17 feet in length. In many early articles on hop culture, it states that a pail was used to carry the twine which ran through the staples. Another way to carry the twine was to make a square bag from hop sacking, or use an old feed bag with an opening in the direction of the top of the stringer to run the twine through. The advantage of this method was that the twine was not separate from the stringer, and did not have to be moved as one walked around the yard.

Two common styles of stringers were those ending in one or more metal eyes, and another that looked like a shepherd's crook. The "metal eye" stringer used a regular pole, with an attached metal top with an eye on the end. These came in many styles since they were often made at home, or were the work of local blacksmiths. One example of this tool I collected came from Fly Creek, and measured 15 inches in length, with two eyes, and the end rounded over to make a final eye. The base of this head is threaded so that it might screw into the pole handle.

Metal stringer from Fly Creek.

The second style of stringer was one of the most graceful and delicate tools used in the hop yards. The "shepherd's crook" stringer was usually made of basswood because it is light and very strong. The crook end was a separate piece that was attached to the pole with either nails or screws. As with other stringers, a series of wire staples would run up the handle to the head. The head had a groove carved on its top, and the basswood head hollowed out to make an opening 3 inches wide. Inside this opening, a wheel was mounted that would turn as the twine passed over it. The end of the crook was also grooved, and had staples to hold the twine in place. This is a much more complicated tool than the simple pole stringer described above. The tool pictured on the following page is 9 feet, 2 inches long, the crook head being the last 26 inches. This stringer was collected on the Morton farm in the Town of Middlefield in Otsego County.

Two other styles of stringers are also known that differ from the types above. I have never talked with anyone that used a stringer, so it is hard to say exactly how these tools were used. Nothing useful has been found in the old literature to describe the methods employed. The first style is a manufactured device that was identified as a stringer when collected and has the words "Patent Pending" on it, but no other information (see following page).

Shepherd's crook hop stringer from Town of Middlefield.

Head of Shepherd's crook hop stringer.

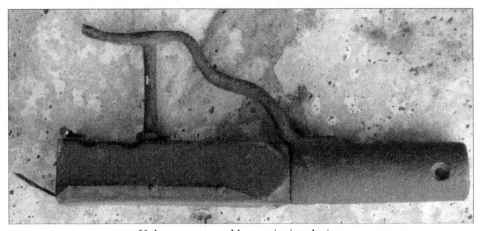

Unknown patented hop-stringing device.

The second style is what I call the "U-stringer" because of the shape of the head. I have seen these tools in the Town of Bridgewater in Onieda County and in the Town of Brookfield in Madison County.

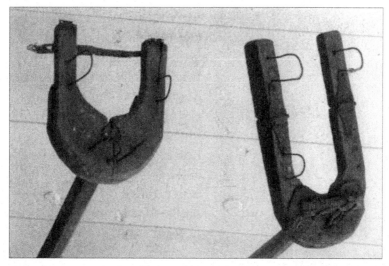

U-shaped hop stringer.

The man that did the stringing was considered a skilled worker. It was a difficult job to properly string a yard, and one that only experienced hands did. The best day to string was a cloudy day, since he would be looking up all the time and the sun would be hard on his eyes. Also, his neck and shoulders would become very tired after a long period of working a stringer. To make matters even worse, the string might break, making it necessary for him to rework all or part of a row.

Closely related to the twining and stringing of the hops is a process called "winding." Once the plant started up the pole or the strings, it was often necessary to continue to direct the hop on the pole or string. As the bine grew it would get too high for a person on the ground to be able to direct the hop by hand. To reach the hop, a long-handled, forked stick was used. This was a natural stick with the bark removed and the ends sharpened. Using the stick, the bine could be redirected around the string or the wire. Not all the bines needed winding, only those that had lost their path. In a wire yard the work was more intensive since the bines had to be trained to run horizontally. So every

Natural forked stick used to train the bines, about 7 in. long.

Using a ladder at Child's hop yard to train the hops on the poles.

few days the yard would have to be checked, and those bines needing it, redirected.

Another interesting way of winding the hops was with a horse. By standing on a horse's back, the hops could be reached until they had grown to a considerable height. A grower from Hubbardsville in Madison County recounted how his mother made a specialty of working the higher bines while standing on the rump of a horse. The men would walk along with her and set the lower, growing bines. This was faster and saved a lot of labor, as the alternative was to use a ladder. "When they are beyond one's reach, use a ladder similar to a fruit ladder." The winding work went on well into the summer. "It is well to look the yard over as often as once a week and attend to the vines that may be off."

Hop yard early in the season, Seward. Notice the hops growing up the strings to the tops of the poles.

# Cultivating the Hops

Grubbing, training, tying, stringing and winding the hops were tasks that often were performed simultaneously in different parts of the yards. Leon's entry for May 8 reads, "Arthur twined hops all day. Still grubbing the hops." And well into June (11-13) the diary records, "Putting hops on twine."

At the same time, in other parts of the yards, cultivation of the hops started as early as April 23. This process would continue until the yards were so thick that it was impossible to work a horse-drawn cultivator through the rows, usually by the end of June. This advice from Jimmy Clark was one of his keys to success as a hop grower. His yards were always famous for their weed-free appearance.

> The cultivator should be used sufficiently often to keep down weeds, and the hills should be dressed with a hoe three or four times. Never let the weeds get the upper hand. As soon as the poles or stakes are set, start the cultivation, three or four times in a row both ways, and keep going over the yard every week until within about two weeks of picking. Whatever may be neglected, don't fail to cultivate, cultivate, cultivate, as that loosens the soil, admits sun and air, releases the plant food, keeps down the weeds, and advances and increases the crop very materially. Late cultivation also helps to bring the hops to burr.

Cultivators used in the hop yards were of several styles. Some cultivators were specifically used in the hop yards while others were more general cultivators. The photo below of four cultivators from the 1995 hop seminar shows the basic types.

The cultivators on each end of the row in the photo were commonly used in the hop yards. They are called "shovel" cultivators, or more commonly "Go-Devils." Several Upstate firms made these cultivators, particularly the Munnsville foundry. Some have the maker's name on the handles, and are often a bright, salmon-red color. These cultivators were very difficult to use and would tire a man out after a day of working with one.

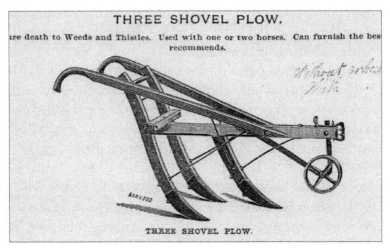

Go-Devil cultivator made in Munnsville.

The second cultivator from the left in the photo is a "duck foot" cultivator. They were so named because the iron cultivators are reminiscent of the webbed foot of a duck. This cultivator featured two adjustable sets of cultivators. If the row was wider, the two arms on the sides could be opened to cultivate a wider row. Munnsville also made these cultivators, as did other firms.

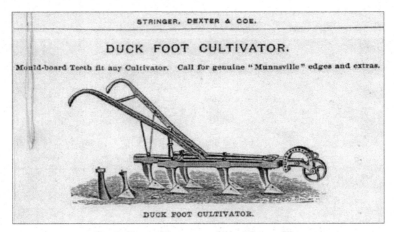

Duck foot cultivator made in Munnsville.

The third cultivator in the photo is a common, all-purpose cultivator. These were used to cultivate almost any crops and to work the hops. They are seldom marked and are quite common.

A companion piece with the cultivator was the hiller. After the hops were cultivated, the loosened soil would be pushed up around the hills with this device. Many people today call these tools "potato hillers," as they were often used for that purpose, but they were also commonly used to hill the hops. They remind one of a winged snowplow. The center of the hiller is a stationary shovel plow, and on each side are adjustable wings. As the hiller moved down the row, it pushed the soil from the row to the edge of the row near the hop hills. The hiller in the photo below is marked, "T.P. Fish, Cedarville NY."

Hop hiller.

After the hop yards were cultivated and the hiller ran through, the next step was to hill the hops by hand. This was done with a hop hoe. Hop hoes differed from the ordinary hoe in that they were much larger. They measured approximately 10 inches long, with a height of about 6 inches. The tops were either flat with curved corners, or arched on either end. Most examples extend into the wooden handle, where a nail through the handle secures them. These hoes are fairly hard to find today, since they were very usable for other farm chores. One informant recounted how they used their hop hoes to scrape the interior of the chicken coops and simply wore the blade down over the years. Using the hoe, each hill had the loosened soil drawn up around the hop, and at the same time fertilizer would be worked into the ground. Usually the hilling was completed by late June. In the Van Patten diary the entry for June 22 reads, "Hilling hops, putting on phosphates and plowing in."

Otsego County hop hoes.

Hoeing a hop yard in early May.

Fertilizing the hops during the growing season was necessary to produce a good crop. Fertilization was simple and the fertilizers varied with each farmer's views. The fertilizers used by Leon were typical of most farmers. In June, phosphates were used.

Phosphates were often either bird or bat manure that was sold for farm use. These were placed around the hill, or broadcast in a general pattern. In both cases the phosphates were worked into the soil. Other fertilizers used in June were mixes of lime, plaster, ashes and hen manure, at the rate of 1 pint per hill. Others used fine ground bone, which was applied to each hill.

The second application of fertilizer was made in July, when the hops were "in the burr." This was very important because it gave the plant food to make the hop. By this point, the plant had developed an extensive system of feeder roots, so it could absorb the fertilizer easily. On July 16 the Van Pattens were putting a combination of hen manure and ashes on the hops. The hen manure was considered very good because it had few weed seeds, and was very high in nutrients for the plant.

When the hops were plowed and cultivated and a yard laid out, I usually discuss the hop hills, but the true hill does not appear until the final hoeing and fertilization in late June and July. Until that time the yard was fairly level. Not until the final grubbing was completed, and the cultivation well underway, did the plants have the soil hilled up around them. Also, this had to wait until the soil was fairly dry, which would be in June. In earlier times in Europe, hops were grown on high hills, and the habit of speaking of "hop hills" continues, even though the practice changed in America.

## Dusting the Hops

In the spring and into the summer, it was necessary to dust the hops to combat mold. In the hop region of New York, the "blue mold" was a major cause of concern for the growers. The mold was known in the 19th century, but devastated the hop region in the early years of the 20th century. From early May until the hops were ready to harvest, the plant was at risk from the disease.

The mold attacked the leaves of the hop, forming white powdery spots that have a circular outline. The mold might totally cover the leaves and form white flowery growth. On the flower, the white spots would cover the developing hop and make it look as if dusted with flour. This would stop the growth of the hop, making it fail to "hop out."

In 1909 and 1910, the blue mold spread over large areas of the hop district. It was carried in the air from yard to yard, and little could be done to stop its rapid devastation.

Growers might have recovered from this onslaught, but the scourge returned with more vigor in 1913 and 1914. In 1913, in some sections of the hop-growing area, nearly two-thirds of the crop went unpicked, since the hops were ruined. This continued outbreak of blue mold was a major cause of the decline of a weakened hop industry in New York State.

Efforts were made to fight the blue mold, but their success was limited. With the aid of Cornell Unversity, experimental stations were started to test methods of dusting the hops. The plants were dusted with a finely ground sulfur. This process was a very disagreeable task, and not always a success. A horse-drawn dusting machine was used. It had a circular hopper with a fan-driven duster that blew the fine, powdered sulfur up a tin chute and out onto the hops. From the pictures of dusting, it is clear that neither people nor horses enjoyed this work. Both men and animals wore protective masks to keep the sulfur out of their noses and throats. The driver also wore protective goggles for his eyes and completely covered himself to keep the sulfur off his body.

Dusting started early in the season, no later than the second tying of the hops. The goal of dusting was to prevent the mold, so an early start was necessary. A second application was needed when the hops were well up the poles, and a final application was made when the hops were well formed before harvest. In some years more applications were needed. If heavy rains came, the sulfur might be washed away, and an additional application was needed.

Dusting was always done first thing in the morning. The ideal day would be one with heavy dew. This would help the sulfur adhere to the plants. A grower that used mules to pull the duster recalled how hard it was to get the mule into the duster because he so disliked the job.

"Teddy" the horse ready to dust. Notice the bag over his muzzle and the dress of the driver. At Cherry Valley, early in the season.

Dusting did work, but it required a lot of time and effort. Mold might still appear even with the dusting, but not as extensively. Also, the cost of dusting was a factor. Cornell estimated that the cost was about $5.25 per acre for sulfur, about $2.00 more for labor, and the duster $75.00. This represented a sizable outlay for a small hop grower.

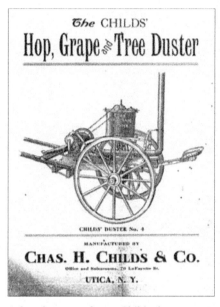

Advertisement for a Childs hop, grape and tree duster.

In addition to dusting, some efforts were made to spray the yards. Again, the use of sulfur was the standard spray material. Special two-wheel carts, with a mounted barrel and extended spray arm would be driven through the yards. I have never seen a picture of a sprayer, but I have seen one in person.

By mid-July the first part of the grower's year would be over. The yards would have been set, cultivated, fertilized, and the hops growing well. Now was the time for the farmer to turn to other pursuits, such as haying. The hops would continue to grow with a minimum of care until harvest.

Heavy rains and windstorms might interrupt this brief respite. Storms might easily knock down the hop poles. If the poles did not break, they were simply reset. Sometimes the poles were broken off, and this required more work. Special conical-shaped iron points were used to reset the poles. These points were driven onto the pole, and the pole pushed into a new hole.

But hopefully, no problems would upset the farmer's world, and a period of calm would precede the harvesting season.

# August and September: Harvest and Processing

As July turned to August, the preparation and anticipation of the hop harvest produced a busy time for the farm families all across Central New York. Farmers looked forward to a successful harvest and a profitable year. However, important chores had to be completed before the farm was ready to harvest and process the hops and ensure a successful harvest.

## Preparing the Hop House

In the off-season the hop house was usually little more than a storage building.[*] Equipment such as hop boxes, the press, cultivators, etc., along with other farm tools and equipment, were stored at various times in the kiln. Then all of these items would have to be removed and the building prepared for the drying and processing of the crop. This work was done primarily by the men on the farm.

Some of the preparation work was as simple as bringing a supply of wood or coal to the kiln. Wood was the main fuel used in drying the hops, and a good supply had to always be at hand. Going along with the drying idea, the stove would be checked and the stovepipes made ready for the harvest. The stovepipes were kept in place throughout the year and only cleaned before each season.

The drying floor had to be inspected to make sure the cloth covering was in good repair. Usually these repairs were anticipated the year before. During the winter months, the women of the house would have the job of sewing the three-foot pieces of kiln cloth together to create a covering for the slatted drying floor. In the earlier days, the kiln cloth was a loosely-woven linen, but by the late 19th century, a coarse burlap material was in common use. Part or all of the kiln cloth might be replaced in a given year. The old cloth was removed, and the new covering stretched tight across the slats and tacked along the edges.

---

[*]The layout and different styles of hop houses are addressed in the Appendix.

# Harvest & Processing

Drying floor of a hop house in Roseboom. The left side was still covered with hop cloth.

To protect the crop that might have the most value, the hop house also had to be cleaned. The dried hops would be in the storage area for more than a month, so it was important to have that area clean. A thermometer was also placed near the furnace room door so that the temperature in the kiln could be monitored. Apparently large mercury thermometers were used, but no extant example is known.

The final task in preparing the hop house for use was the building of a platform at the door to the drying floor. Hops were brought from the hop yards in large sacks. These were removed from the wagon onto the ramps, and then taken into the drying floor and spread evenly. Some hop houses had special small ledges attached to the side of the building to help hold the floor of the platform (see following page). In a few cases a dirt berm was built to make it easier to get the hops into the kiln. The berm would make for a lower platform, and make the passing of the hop sacks onto the platform easier. When all these chores were complete, the farmer could turn his attention to preparing the tools needed for the harvest.

During the off-season, the equipment used in harvesting was also commonly kept in the hop house. As harvest time neared, the job of preparing the hop boxes and other tools was a pressing concern. Hop boxes might be stored in the furnace room if a large exterior door was available and the room was big enough. If the furnace room was not accessible, the boxes might be placed upstairs in the storage room. I have seen several hop houses in the Upstate area where the stored hop boxes are still in place in these two rooms, as if waiting to be used in next year's harvest.

Step-up hop house with temporary scaffolding, Hubbardsville.

Hop boxes lasted for a good many years and were simple to make and repair. Most of the boxes I have seen show some sign of repair. A broken board might be replaced, or the handles might need strengthening or a new set made. In most hop boxes a repair will show because of several characteristics. The nails used might change from square cut nails to wire nails. The replaced board might not be painted. A lack of names and dates left by the pickers would reflect the fact that the board was new. On some farms a number was painted or stenciled on the box to keep track of how many hops were picked into that box. If these numbers needed touching up or repainting that was also done before the boxes went to the yards. Once the repairs were done, the boxes might be placed to the side around the hop house, or taken directly to the field in anticipation of the picking. To take the boxes to the yards, a modified work wagon was used. A special frame was built that extended about 3 feet out from the sides of the wagon bed. This made it possible to carry as many as nine boxes.

Supplies had to be purchased for the upcoming season. A number of items were needed, such as sulfur sticks and hop sacking for baling. Local merchants would have these supplies on hand, and placed extensive adds in the area papers to attract prospective buyers.

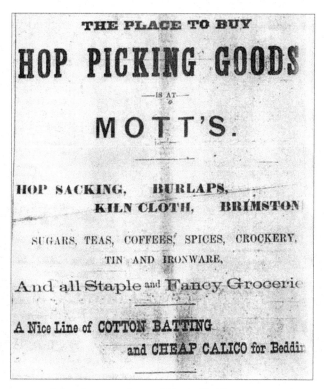

Advertisement from a Waterville Newspaper.

On some farms the owner saw to the purchasing of insurance on the hop house and the hop crop. Fire was the big danger, as the hop stoves were burned very hot and the dried hops would make excellent kindling. These policies were for short terms, about sixty days.

Some farmers also needed to go to the local print shop to get tickets for the upcoming harvest (see following page). Hop tickets were issued to the pickers according to how much they picked. They came in a variety of denominations, from a quarter box to a single box. At the end of the picking season, these tickets would be redeemed at the going rate per box.

Another way to record the amount of hops picked was with a printed punch card. These were punched once for each box or part of a box picked. I have been told that metal tickets were also in use, but I have never seen any from the upstate area.

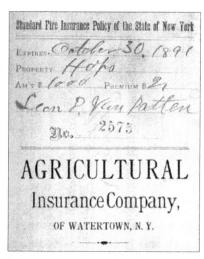

Leon Van Patten's hop insurance policy, Middlefield Center, 1891.

Probably the simplest way to keep a record of the boxes picked was for the yard boss to keep a tally under each picker's name.

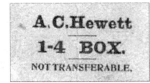  

Hop tickets.

Punch card used in Madison. Each number represents one box.

It is easy to see that the time from the last cultivation to the harvest was a period of much activity for the farmer. One did not simply rock on the porch and listen to the hops grow, a lot of preparation was needed. Also, other farm work was going on at this time, such as making hay, which was very important to the dairy side of the diverse farming of that time. In addition, a second level of preparation was also underway.

## Preparing for the Pickers

The women of the farm were responsible for getting the farm ready for the hop pickers, both day-pickers and those that were boarded for longer terms. The number of boarding pickers determined how much space was devoted to sleeping quarters, so various dormitory areas had to be made ready.

The women pickers usually slept in the main house. There were often large, open areas above the kitchen or woodshed at the rear of the farmhouse. In many of the old farmhouses, a separate stairway would lead up to this large sleeping area. The men might be housed in special rooms in the barn, over a summer kitchen, in the hop house, or, as on some farms, in separate dormitory buildings. In all cases a strict segregation of the sexes was the rule.

In all these areas, the women had to clean and prepare the bedding for the pickers. Beds were brought from storage, and assembled and roped. Mattresses and pillows needed new stuffing of straw, corn husks or excelsior. Usually this bedding was put away and used only during the picking season.

Assembled hop picker's bed waiting to be roped, Madison.

The rest of the house had to be prepared for the work of feeding the pickers during the harvest. Carpets were rolled up or covered over to protect them from the heavy use of so many people moving through the house. The long hop tables and benches had to be taken from storage and cleaned up in anticipation of a busy season. Other details that needed attention were sanitation facilities and arrangements for the pickers to wash after work and before meals.

The hardest work the women faced was the task of food preparation for the onslaught of the famished pickers. Food preparation was an absolute necessity for a successful harvest. As much food as possible would be made beforehand to cut down on the amount of labor in the kitchen during harvest. All boarded pickers were served three meals a day. In addition, the local day-pickers might also be given a noon meal, though many brought their lunch pails with them. In hop-picking times, the big meal of the day was dinner, which was served at noon. The day started with a large breakfast in the morning, and ended with a light supper in the evening.

The quality of the food provided was a point of pride with the farm wife. Meals were hearty, basic fare, but nothing fancy. Usually the same staples of the farm diet were served to the pickers, the only difference being the amount of food served. Those that worked in the hop yards always seemed to recall the huge appetites that the pickers had. Many a picker remembered the quality of the food served. In a hop poem from the 1880s, the author praises the cooking.

> It would take volumes of books,
> To tell all the praise due our cooks.
> While we eagerly devour the food,
> The general exclamation is Oh! How good!

The poetry may not be the best, but the sentiment about the cooking is clear.

where the most work is done

Many believed that the pungent odor of the hops was the cause of such good appetites. Probably the hard, physical work in the hot, late-summer sun was a better explanation. The workday started as soon as the hops were dry enough to start picking, and supper was not until six o'clock, so the day was long.

To prepare for these appetites, the farm wife had to start preparations weeks before the picking. Mary Ferguson of Oneonta gave the following account of the work involved in feeding the pickers.

> My people raised hops for many years. We wouldn't have been considered one of the large hop-growers. During hop-picking season, we had about 25 pickers for three weeks. Many of them were friends, some roomed at home and boarded with us; others slept in a house on the farm.
>
> From the time I was 12 or 13 years of age, I helped my mother in the house. We did all the cooking. Mother made at least 8 loaves of bread each day. We baked pudding or pies each day. We always had cake for supper. We bought meat (beef) by the quarter. There were no refrigerators but we had a cool cellar. We always planned to

have a large garden with plenty of sweet corn, tomatoes, cucumbers, cabbage, and other vegetables. Also we had a quantity of fruit.

During that period breakfast was served at 6 a.m., dinner at 12, and supper at 6 p.m.

People who picked hops had a ravenous appetite. We prepared accordingly.

Father dried hops at night and he always peeled one bushel of potatoes each night. Otherwise, mother and I set the table, washed the dishes, baked, prepared vegetables, and what else was needed.

It is notable that the farm was very self-sufficient in providing most of the food needed during hop picking, and that the man of the house shared in the food preparation at night in the hop kiln.

On many farms in the days immediately before picking, the women might prepare ahead by backing items that would stay fresh until the picking started. Cookies and doughnuts were made in large quantities and stored for the pickers. Clara Martin of Milford remembered as a girl helping her mother make molasses cookies in preparation for the season. The cookies, once cooled, were placed in wooden, 100-pound flour barrels for storage until needed. Since the usual items of the farm diet were used during hop season, the ever-popular sugar cookies and gingersnaps were commonly made. Plain, fried doughnuts also might be made ahead and stored until needed. All these food items were a mainstay of breakfast, and sometimes served with supper.

The morning meal had to be substantial, as the day started early and the work was hard. It was common to serve two hardy foods, such as oatmeal, fried potatoes, hash, creamed codfish, fried salt pork, or dried beef with gravy. All this was supplemented with bread and butter, coffee, cookies or doughnuts.

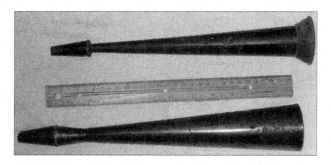

At noon the tin dinner horn or bell called the work force from the yards into the farmhouse for the main meal of the day, dinner. Usually a meat course, either beef or chicken, one or two fresh vegetables, potatoes, coffee and a pie for dessert were served.

The quantity of food at noon was important, as the leftovers would be the basis for supper at six that evening. Cold meat and vegetables would be the center of this last meal of the day. The universal potato was also served. Commonly at supper some type of sauce, such as apple or rhubarb, was put out. And finally a plain cake, one with no icing or filling, and cookies would conclude the meal. With a cooking schedule like this, the farm wife found the hop season to be a very busy and strenuous time.

The importance of the food during hop-picking season must be stressed. Pickers became attached to a farm that set a good table. It was not uncommon to be warned that a certain farm served only one dish at dinner, such as codfish gravy, and should be avoided. In many hop houses and on hop equipment, you can find the names of pickers and the years they picked at that farm. Often several dates will follow, which reflects the loyalty a picker had for a farm that fed its boarders well.

## The Harvest

Throughout the hop region of Central New York, the harvest started in mid-August and extended, in some areas, into the middle of September. It is hard to give very exact dates, since so many factors influenced the time of harvest. Weather had a great effect on when the hops were ripe. A hot summer might bring the hops on faster. The amount of moisture received would also affect the harvest. Other factors, such as the farm being located in a valley or on a hill, caused the hops to ripen sooner or later. Many yards also had several varieties of hops that ripened at different times, so the picking might be extended over a longer time period.

Growers had to be able to determine when the hops were ready to pick. The appearance of the "strobile," commonly called a hop, was a good way to tell if the crop was ripe. The hop cone had to be tight and drawn together at the tip. The hop also had to have a hard and solid feel when grasped.

The most observable characteristic of the hop is its color, which should be a light green. Inside the hop, the seeds should be hard and brown, and a good supply of yellow lupulin should be evident.

The quality that was most remembered by former hop-pickers was the odor of the hop. Those that picked hops called this smell a "ripe" smell. By rubbing a hop between your fingers, this very characteristic smell will immediately be sensed. The hop will also produce a sticky feeling on your hand, which is also a sign of a mature hop.

The final test was to shake the clusters of hops and listen to how they sound. Shaking the pole and hearing a rustling sound indicated hops that were ready to harvest.

Most agreed that hops picked a little early were more desirable than those that were picked when too mature. The trick was to start picking just a little before the hops were at peak ripeness. As the harvest progressed the degree of ripeness would vary, but the early hops were considered the best in the 19th century. Brewers stressed the light green color in buying hops, so the early hops were the most desirable.

A few years ago when I was interviewing Hank Springett, he told me that his mother always said the he was born in the middle of a hop yard. Hank was born in 1917 in Hubbardsville in Madison County. Today when you drive through that area of Madison County in late July and early August, the landscape has few, if any, reminders of the way it looked during the hop-growing years. It is hard to envision a land as described by James Fenimore Cooper in <u>Reminiscences of Mid-Victorian Cooperstown</u>: "… the whole countryside was one great hop yard, and beautiful."

## Hop Boxes

Hops were picked in a variety of ways, using several types of containers. In the early years of the 19th century, hops were picked into baskets, bins or cribs. This method of picking was common in England and probably brought to the United States at an early date. In England the use of bins continued throughout the 19th century, and was the preferred method. But by the middle of the 19th century, the hop box became the universal container in New York State. Boxes were usually supplemented by various size baskets for side-picking. At the turn of the 20th century, special large, wooden baskets and metal frames to hold a hop sack came into use in many areas. No matter what the container, the same basic team approach was used in the yards.

Usually a team of four pickers and a box tender made up a set. In earlier years, as many as six to eight pickers might work a single "bin." Pickers were most often women and children, but men also helped. The gender of the pickers was not as important as the quality and the quantity of the picking. The hop had to be picked clean; no leaves or stems would be allowed into the box. An average worker was expected to pick about 15 bushels of hops per day. A good picker might pick 25 bushels or more. When you look at the size of a hop and the amount picked, 15 bushels looks like a very large quantity of hops. To accomplish that task one had to keep on task and keep moving.

The February 1836 issue of *The Cultivator*, a popular agricultural publication of the period, describes the hop bin as follows:

> It is a frame 8ft. long by 3ft. wide supported by three feet tall legs at the corners. On each end of the frame would be a support to hold a pole that ran the length of the frame and extended several feet past the end of the bin on each end. These extensions were important because they were rests for hop poles during picking. The bin itself was covered by a cloth, probably a "burlap type," hooked on the long sides of the bin. Around the bin from six to eight pickers might work. In addition a man was needed to bring the poles to the bin for picking.

In an 1839 article in *The Cultivator*, it was stressed that only two poles at a time should be brought to the set, and never should a pole be pulled more than one hour ahead of the picking. Hops wilt very rapidly once pulled, which would have made picking more difficult and caused the deterioration of the hop's quality.

So far no existing hop bins are known. In my exploration of hop houses I once felt that I had found a hop bin. A very old hop house in the Town of Richfield was the scene of this near sighting. The building was in a very fragile state. The kiln was separating from the rest of the hop house. The drying floor appeared to be unsafe to walk on and, in places, looked ready to collapse. The drying floor was entered through the exterior door used to bring the hops in for drying. In a corner of the drying areas Dave Petri and I saw what looked like two collapsible frames. I immediately jumped to the conclusion that they were bins. The question was how to get these objects? With the help of a very long stick, we carefully retrieved both artifacts from the corner of the kiln. Unfortunately the finds were not hop bins, but rather two folding beds. Both had good age on them, and are interesting examples of picker's beds. The use of hop boxes came in so early that it would be most unlikely that any hop bins will be found. Since most hop artifacts are from the second half of the 19th century or later, these very early tools would be extremely rare.

I found it interesting the way the hops were removed from the bins. The hop was picked directly into the cloth-covered bin. Once the bin was full, a man called a "sacker" would come and scoop the hops out of the bin into a large sack for transport to the hop house. The cloth on the frame remained on the frame, and the process of picking could resume. Later, the hop sack became standardized to hold 8 bushels of hops, which was one box. The size of these early sacks is not known and might not have been standardized. It is also not clear if the pickers were paid by the bin or by the weight of the hops. Later the practice was generally to pay by the box. Since the measurement of the bin might have been less exact, the hops could well have been weighed in the field. Some very early hand-forged field scales have been found in hop houses, and might have been used to weigh the hops.

Group of pickers with the hop sacker holding a sack of hops on the left.

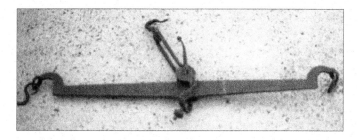

Wrought-iron yard scale from the Green farm, Hartwick.

As noted before, the bin system continued in England for many years, but never seemed to become popular in America. The concept, however, did not die out completely. In 1899, Levi M. Myers patented a hop bin that is very much based on the early model (see following page). He wanted something that might be easily moved around the yard and easily stored in the off-season. In the patent, the ends were hinged so the bin would fold up easily. This was a western state patent, and no evidence has been found to show that this idea was ever tried in our state.

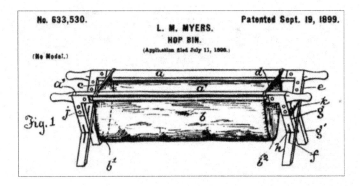

Patent drawing for a Myers hop bin.

Well before the middle of the 19th century, the use of hop boxes was the most common method of picking in the hop counties of Central New York. Hop boxes came in several styles. The most common was the large single box with four partitions inside. It is claimed that Morris Terry of Waterville was the first to partition a box into four compartments. This fact was mentioned in W.A. Lawrence's article on New York hop growing in Ezra Meeker's book, <u>Hop Culture in the United States</u>, published in 1883. By that date hop boxes had become fairly standardized in our state. Each of the four compartments was approximately 36 inches in length, 18 inches wide, and 26 inches deep. These figures are based on an 1876 law that stipulated the standard box size. From my observation of hop boxes, I would say that many variations of size existed, and the standard was not always followed.

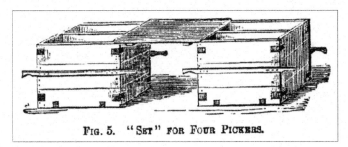

Hop box with an apron between the two halves.

The preferred wood for a box was basswood of thin stock, because of its availability, strength and light weight. In most old boxes the central divider in the box was fixed, while the dividers that split each half of the box were removable. The box corners were reinforced to further strengthen the box.

Hop boxes had to be portable. As one area in the yard was picked, the box would be moved to the next picking site. To make the boxes easily portable, handles were placed on the long sides of the box. They might simply be a board, 1 inch by 4 inches, shaped on the end, and nailed to the box. In some cases bracket-holders of wood were attached to the box, and a flat board extended through them to become a handle. The last type of handle was one that would swing up or down for use as a carrying handle, and out of the way when not in use.

Hop boxes simply sat flat on the ground with no special base. They were commonly painted red, but over the years many have lost their color. If a grower had several boxes in use, he might also place a large number on the box. The inside of the box was not painted. I have seen some boxes with inches marked in a bright yellow on the inside, or a line that would mark a full box.

In New York State the hop box had straight sides. In the West, the box was often narrower at the bottom, and angled out toward the top of the box. I have never seen any box of this type in the Upstate hop country.

Above: Group of pickers around a standard hop box.
Right: Lug pole support.

At the end of the box would be a holder to support the lug pole and sunshade. On many of the boxes I have observed, the holders are made by angling together two boards. They are cut with a bevel on one edge, and come together more at the bottom of the box. This makes it possible to cut an upright at the same angle and bevel its outsides to fit snugly into the holder. The upright has a hole in it about 2 feet above the top of the box. Into this hole will slip a long pole, the lug pole. This is the rest for the hop poles during picking. Finally, the top of the upright would support the sunshade.

Hop pole resting on a lug pole and a set of pickers, Seward.

Hops were picked at one of the hottest times of the year. The sun in late August and early September is still very high in the sky and very hot, thus the sunshade was needed for protection from the sun in the yards. From old hop-picking images it is clear that the quality of these shades varied greatly. Some are simple frames, with what look like large sheets of bed ticking draped over them. Most are more substantial and well made. One of the problems with the permanent shade was that it could not be moved as the sun crossed the sky. To correct this problem efforts were made to come up with a movable shade. In 1883, Jacob Eagle Jr. of Sharon Center patented an umbrella-like device to screen the moving sun. In his application he stated his shade would "afford a simple, efficient and very comfortable protection for the pickers from the sun, that may be readily shifted around as the sun advances." The shade was attached to the box and had a rotating joint that could be moved as needed.

Patented awning from Sharon Center.

Set of pickers under a hop shade with the box tender to the left, Hartwick.

From this basic box two variations are sometimes seen. The next most common style was a box with only two divisions in it. This was simply one half of a large box. Sometimes this was made from a full box by cutting it in half. In a double hop box setup, the hops were picked differently than a single box. Between the two boxes, a wooden platform with a gallery around its edge was placed. This was known as an "apron." (See drawing on page 62.) The hops cut from the poles were placed on the apron and picked from there into the boxes. In a full box the hops were picked directly from the pole resting on a lug pole over the box.

The last type I have seen is a simple box that would hold one sack of hops. These were simply a box without handles, and no supports for a shade. These were once quite common, but very few are now found.

The standardized hop box had an important impact on the payment of pickers. Each of the four compartments was called a "box" and held about 8 bushels of hops. Each picker put her hops into the box that she was assigned for the day. Since the picking was done by boxes, it was easy to pay by the box and not by weight. Certainly by the middle of the 19th century, the practice of paying by the box had become accepted across the hop region.

Hop boxes may still be found across the former hop-growing counties, but they are becoming increasingly rare. Once the hops went out, the boxes had few practical uses. I have been told many times that they were broken up and used as firewood or for other uses around the farm. They also made very good storage boxes for potatoes. The box was placed in the cellar on a dirt floor and the potatoes placed inside. It would only take a few years for the bottom of a box to become rotten, and the ruin of the hop box would be well underway. The boxes I have seen were all in barns or old hop houses. Since the boxes were stored in the off-season in the hop houses, this would be a good place to find them today. I was in a hop house near Cazenovia, and in the furnace room, neatly stacked, were a large number of boxes and sunshades. After the last use of the boxes they were put away and never removed again. Usually boxes are on the second floor in the hop storage area, again a place for off-season storage.

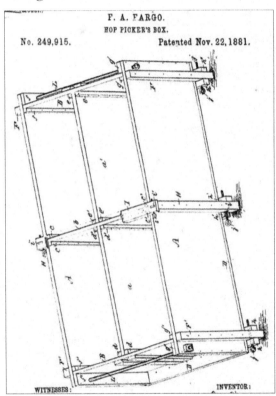

Patent drawing for a Fargo hop box.

Since so much room would be required to store hop boxes, efforts were made to make a box that might break down and take up less storage space. In 1881, Frederick Fargo of Pine Woods in Madison County patented a collapsible hop box. He used tie-rods with threaded ends and bolts to hold the box together. These might be easily unscrewed and removed when not in use, and then stored in a disassembled fashion. This idea never became popular, and no examples of this type of box are known.

It is hard to tell the age of a hop box for several reasons. They were very simple in construction and often repaired over the years. Most of the boxes I have seen did not appear to be very old. Usually they look like something made around 1900. On some boxes are the names, dates and hometowns of pickers carved into the sides of the box. A box that I found near Westville has the earliest date I have seen on a box, 1872.

A hop box was like the bin in the way that the hops were removed from the box. The technique of hand-scooping the hops from the box into a sack was the same as used in earlier times. In a few cases, the hops were scooped out of the box with the help of a large flat coal shovel, but most did it by hand. It would have seemed that a better method of removing the hops from the box might have been developed, but nothing new became popular. (See page 86 for more details about this process.)

An effort was made by William Crandall of Deansville in Oneida County to improve the sacking process. He patented a frame with burlap sacking stretched over it. It looked like a box, but the sides were removable. When the box was full, the sides would slide out, and a full sack of hops would be ready to tie and remove to the kiln. I have never seen a box like this or heard anyone tell about it, so I assume that it never gained any popularity.

In addition to hop boxes, a variety of hop baskets were used in the harvest. By studying the old images of hop pickers, you can observe several styles of baskets being used. Some of the baskets were made of willow, loosely woven in a round shape of about 3 feet in height. A basket like this might hold about 3 or 4 bushels of hops. Others were made of hickory or other split materials. These baskets came in many sizes, and also varied in shape. Since they were used for side-picking, they were not designed to hold a large amount of hops. "Side-picking" is when a picker picked the hops from the lower part of the pole resting on the lug pole. Many times this was the work of children and supplemented the mother's picking. Today almost any large basket found in the hop-growing areas is referred to as a hop basket, but that might not always be the case. Only those baskets with a hop history might honestly be called hop baskets. In the 19th century large baskets had many uses, and size alone does not make a hop basket.

The end view of a Crandall hop box showing the burlap lining.

Small basket from the Neff farm, Madison.

Hop basket, 18 in. diameter, Fly Creek.

In the 1890s and the early decades of the 20th century, some specialized hop baskets came into use. The first was made of wooden slats and held together by a wire frame. It had two bail-type handles and a round wooden bottom. These baskets were first made in the West, and came into use in the East. They are fairly large, holding 8 bushels of hops. The example I have collected came from the Oneida Chief Hop Farm near Bridgewater, and has the name stenciled on the slats.

A second type of basket was a metal form that would hold a hop sack. These were very mobile and did away with the need to remove the hops from the basket, as they were picked directly into the hop sack. The only place I have seen any of these frames was the Moakler farm in Middlefield Township, Otsego County.

Hop basket made of of wooden slats with a wire frame.

Metal form with hop sack.

## The Hop Pickers

The hop box was the focal point of the harvest. The boxes were placed in the yard four rows deep, with four rows on each side of the box. Using this method, 64 hills would be harvested before moving the box to a new site. Since the boxes were movable, the set could be easily moved after picking a square.

The four women that usually picked at any given box had no special tools that they used, but they did have special clothing that helped them overcome difficulties in the yard. Hops had a variety of bugs that made the bines their home. Women always kept tightly buttoned-up when working in the yards to keep bugs from getting under their clothing.

Hops tend to be very itchy and irritating to the skin if no covering is worn. The pickers often made special gloves for hop picking. Before the season, they would remove the fingers from several pairs of gloves. Only the fingers would be exposed to the hop leaves and bines, thus avoiding most discomfort. The pickers sometimes also wore special cloth cuffs. These went up the forearm and gave added protection from the hops.

Hop picker's sleeves owned by Clara Thorn Martin, from the Thorn hop yard pictured on the following page.

Sometimes the women pickers had long smocks that covered their dresses. More commonly, just a large, long apron was sufficient cover. I assume these were worn as protection from the sticky resin that is found in the hop.

The last special item of clothing was the straw hat. These were often very large, and very much needed. Since the picking was done in late summer, the mid-day sun would have been very brutal—thus the need for protection.

All of these items of clothing are seen in the hop-picking images. A close study of these images shows how tightly buttoned-up the women were, and the amount of covering needed for the job.

Group of pickers at the Thorn hop yard, Milford.

With the proper apparel, the picker was now ready to start picking. As with any type of picking, a certain technique was required. The best method of picking is to put your fingers between the hops in the bunch, and not around it. Your hand becomes a comb, and the hop is pulled free from the stem. This way the number of leaves that get in the box is limited, and the hops are picked separate and not joined together by the stems. This produces a clean picking, one with no leaves or stems, only hops. The cleaner the picking, the higher the quality of the hops, and the higher the price one might receive for the crop.

## The Box Tender

Box tenders did not have any special clothing. At times they might use gloves, but many did not. The box tender had several important duties that ensured a smooth picking. His hardest job was pulling the poles or, in a wire yard, cutting the bines for the pickers. He also had to bring the poles or bines to the box and keep the pickers well supplied. The work area also had to be kept clean. The bines, after being picked, had to be placed in a pile for later removal. All the poles were neatly placed in a pile on the ground to be stacked later. Also, when the hops were sacked, the box tender helped by holding the sack while it was being filled.

A box tender might have more than one set to care for, which made this a very demanding job. With 64 hills to each set, the number of poles pulled and handled in a day was considerable. Remember, each pole was about 20 feet tall or more. At harvest they were covered with the hop bines and had a large accumulation of hops on the top of the pole. Each pole had to be brought over to the box. This could actually be dangerous work. If the day was windy, the pole might be caught in a gust and come down hard to the ground. I have an account of this very thing happening, with the result being the death of a picker who was struck by the pole. But generally, the poles did not constitute any danger if properly handled.

## Hop Knives

The box tender, in addition to his responsibilities to pull the poles, had to cut the bines and, in some cases, remove the hops from the poles. These tasks were accomplished with the aid of a hop knife.

Any long-bladed knife with a fairly heavy blade might have been used in the hop yards. I have seen large butcher knives in hop kits that served the purpose, but three special types of knives were used specifically in the hop yards.

The first hop knife was used to cut off the bine about 10 inches from the ground. The bines had to be cut before the pole was pulled. The same knife might also be used in a wire yard to cut the string holding the hops. These knives had long handles, usually anywhere from 18 to 24 inches or more in length. This was necessary so one did not have to bend over to cut the bine. The blade of the knife varied in length and curvature. On some examples the blades are only about 2 inches in length, in others as much as 4 inches.

All of these knives I have seen were handmade to meet the specific needs of the owner. On some, the blades are simply at a right angle to the handle, with only a little curve. In others the blade had a hook shape. Since these were handmade, the blades were often made from old files. On many examples the worn-down file marks can still be seen. Sometimes a small loop of leather or rope is at the end of the handle to secure the tool around the user's wrist. One of the most interesting examples I have found is made from an axe handle, with a handmade, hooked blade of 4 inches.

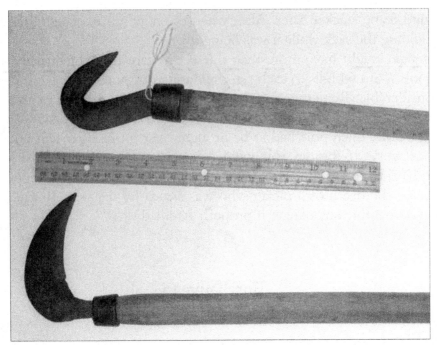

Hop knives made from files, Madison County.

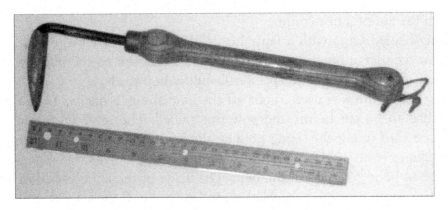

Short-handled hop knife for cutting bines.

    The second style of hop knife was used to cut the hops from the poles. These knives were also homemade, or made by local blacksmiths. They also represent an early example of recycling. When the blade of a scythe was worn out, the end of the blade was cut off. Using the last 25 inches or so of the blade, about 15 inches of the thin metal would be cut out. The remaining heavy, metal top of the blade would then be turned toward the point of the blade to produce a handle. The last 10 inches of the blade then became the blade of the hop knife. In many of the old hop-picking images, the men in the photos proudly display these knives. Often they have blunt or cut-off ends, since old and broken blades were used. When the hops were cut from the poles to be picked from the apron, this was the tool used for the job. It was also used to strip the bines from the poles.

It is interesting how people have found these knives over the years. Almost any one that has this type of a knife will have a story of finding their knife in an old hop house. Many times they were placed on the rafters in the buildings, where they would be discovered many years later. I searched and searched in hop houses for a knife, but was frustrated for many years. One day I was helping a friend remove a hop box from a kiln in the Town of Sharon. The box had been stored standing on its narrow end, with its bottom out. Actually it did not even look like a hop box in that position. We carried the box outside into the good light. As I was looking it over, I noticed a hop knife lodged in a corner of the box. That was a very happy day, as I then had a story about finding a hop knife in an old hop house.

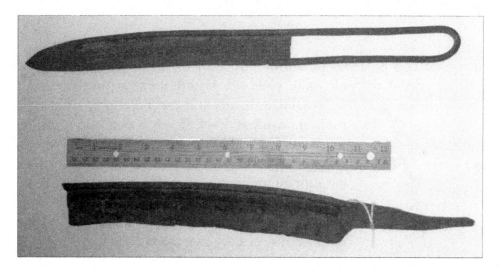

Hop knives from Lawyersville (top) and Breakabeen (bottom) in Schoharie County.

The last style was the least common of the hop knives. I first noticed this tool in the old photos. Over the years I have looked with care for examples, but have only found one. These tools had a straight, wooden handle of 12 inches or more in length. The blade was also close to 12 inches, with a slight curve that came to a point. In my example, the blade is 11 inches long, and about $2^{1/4}$ inches wide. It is handmade, and sharpened along the curved lower edge. My example came from the Hamilton area in Madison County. I have never been told about the use of these knives by anyone, but from the old photos I know they were used.

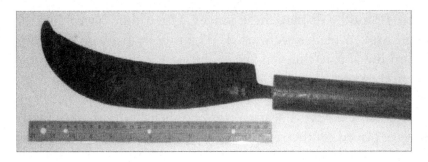

Above: Hop knife from Madison County.
Right: Unknown box tender holding a similar hop knife.

# Pole Pullers

Pulling the hop pole. From the collection of Mrs. Homer Martin.

The pulling of a hop pole was most often done by hand. To do this, you place your feet on either side of the pole, with your legs slightly bent at the knees. Then you wrap your arms around the pole and, with a push-pull motion, work it loose from the ground. At the same time, you lift with your legs, and hopefully the pole will come out of the ground. Obviously, factors like the amount of moisture or rocks in the ground, and how deep the pole was buried would affect the success of pulling the pole. If the pole was a "sticker," it then would have been necessary to use a pole puller.

Pole pullers came in a variety of styles. Some of the devices were patented, but most were not. Many of the small foundries made tools for the hop growers. Other pole pullers were made by local blacksmiths, or on the farm.

The simplest pole puller was a length of chain with a circular link of about 2 inches in diameter on the end. This circular link would be placed over a sturdy pole about 6 feet long. The chain would be wound around the hop pole at a height of 3 or 4 feet. The chain would then be secured to the pole by a nail, or by running it through the circular link. Then, using the pole as a lever, the hop pole could be raised and loosened from the ground. I have been told about the use of this method and have found mention of it in the old literature, but have only seen one chain and loop that might have been used in a hop yard. Since a chain like this would have many farm uses, it was more of an all-purpose tool than a specialized hop tool.

The most common puller was known as a "hop dog" or "L-shaped" puller. They came in several slightly different styles and sizes, but all had the basic "L" shape. They were made of two pieces, joined at a pivot point slightly above the angle of the L. Above the pivot, both pieces had a loop to hold the strap that attached to the puller. On the other side was the 90 degree angle and the teeth of the puller. I assume they are called "hop dogs" because of the small tooth-like projections that clamped into the pole to hold it securely, reminiscent of a dog's jaw. On some examples, as many as sixteen small teeth are found on each side. The jaws of the puller, when open, show a slight curvature to help grip the pole. The standard size for this puller was about $7^{1/2}$ inches long, with the jaws of the puller being about $4^{1/2}$ inches long.

Out of the factory they were painted black, but many over the years have lost their color. I have seen one that was a handmade copy of the foundry product, but slightly larger.

In my collection I have a very interesting repaired example. The repaired puller is half factory-made and half hand-forged.

Very few of these pullers were marked by the maker. The only marked example I have ever seen is from Deansboro, marked with the name of "F.N. Satterlee."

Two views of a common hop dog or L-shaped hop pole puller.

Hop dogs always had some type of a strap attached to the two loops. The straps were of a variety of materials: leather straps 1 inch wide or more; heavy canvas material with a leather end that attached to the dog; hop cloth (a rough burlap material); or other less common fabrics. Usually the strap was long enough to extend around one's neck and hold the dog at about waist height, or a little lower. In some cases a harness that went over both arms was used, but does not appear to have been as popular as the simple strap over the neck.

Satterlee hop dog and canvas strap, Oneida Chief Hop Farm, Bridgewater.

I have used these pullers and they are a very effective way to start a pole or to completely draw it from the ground. You place the hop dog around your neck and stand before the pole as I explained earlier, but you also attach the jaws of the dog to the pole. Then you securely wrap your arms around the pole and straighten your knees, so that the power of the pull comes from the legs and not the back. The teeth will grip into the pole and with luck it will come up.

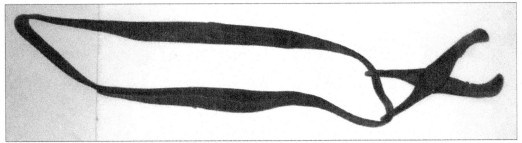

Hop dog with leather strap, Esperance.

The second type of pole puller is what I call the "ice tong" puller. These were simpler than the hop dog. There were two arms with a riveted pivot point. The top of the puller was very similar to the hop dog, with an elongated loop on the end to hold a strap. The bottom of the arm curved to a single point, reminding one of a small pair of ice tongs. The tool measured about 10 inches in length, with the tongs being about $4^{1/2}$ inches long. I have seen several different types of loops at the top of these pullers that might reflect that several foundries made them. Most of the ice tong pullers were manufactured, but handmade examples are known. One of my favorite pole pullers in my collection is a small ice tong type, about $7^{1/2}$ inches. This piece is totally hand-forged and more pointed at the end than the manufactured types.

Wrought-iron ice tong puller from Madison County.

Like the hop dog, a strap would attach to the puller and be draped over one's neck when being used. When using this tool, it was important to have the points of the tool firmly attached to the pole. The method of use would be the same as the hop dog.

Hand-forged ice tong puller with strap, Otsego County.

Most examples of ice tong pullers are from the Towns of Cherry Valley and Middlefield in Otsego County. My hand-forged example came from Madison County in Hamilton. No marked examples are known at this time.

The third type of puller is known as the "W-shaped" puller. Like the letter "W," this tool has two openings with a peak in the middle. Very few of these pullers have been found. The examples that have come to light have all been found in the Cherry Valley and Middlefield areas. In the manufactured examples, the cast-iron puller has two half-circular sides, joined in the middle. Near the end of each half-circle is a small nib. This was to secure the pole when using the puller. A chain of two or three links is attached to a hole in the middle of the puller. This is topped with a round link to hold the attached strap. In all the examples I have seen, one side of the puller is larger than the other. This would be a way to adjust for larger or smaller poles.

Several years ago, at an auction in the hamlet of Middlefield, I purchased two hand-forged examples. Obviously the maker had seen the manufactured item and had made his own. The one difference in the handmade example is that the little nibs were left out, creating a totally smooth surface. I do not think that these would have worked as well, since they might slip when pressure was applied. I have never seen any of these pullers in old photos, but have been told by folks that they were used in the yards. No marked examples are known.

W-shaped puller, Middlefield.

# Harvest & Processing

Wrought-iron W-shaped puller, Middlefield.

 Yet another type was the "Y-shaped" puller. A piece of iron was forged into a "Y" shape, with an opening of about 4 inches. On the inside of the opening a series of teeth were cut to grip the pole. Some type of a wooden handle was attached to this tool, but no example with a handle is known. The tool would be like a lever, and start the pole out of the ground. All examples that are known are handmade, and they all differ slightly in size and shape. Some examples have been found in fence rows in old hop yards, but no one has ever identified the tool as a pole puller.

 Earlier I discussed this tool as a support used in removing the bark from hop poles. I am still not sure of its exact use, and indeed it might have had several uses. Hopefully some day a complete example will emerge and someone will be found that can positively identify it.

 In 1879, B.G. Chapman of Clayville in Oneida County patented a machine for pulling hop poles. Over the years I have collected two of these patented pole pullers. The first puller I found came from South Edmeston, and the second from the Hamilton area. They were both the same, but the second had the original forged-iron puller, while the other had a replacement hop dog.

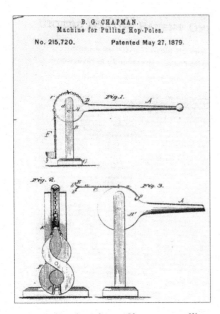

Patent drawing for a Chapman pulling machine.

The basic machine is made of wood, with a base and an upright that supports a lever with a large circular head. Attached to the lever where the circular head begins, is a chain that is held in a groove in the circular head. The chain holds a forged-iron device that actually attaches to the pole and lifts it. The wood on these machines was originally given a red wash, marked with "Chapman" and the date of patent stenciled on the upright.

I have tried my puller and it is a very efficient tool. By placing the foot of the machine near the hop pole and lifting the lever arm upward, the iron puller is made ready to attach to the pole. The puller opens much like the hop dog and is clasped to the pole. By pressing down on the lever arm, the puller is tightened and the pole is lifted. If another pull is needed, one can simply lower the lever arm, and the puller will reset and be ready for another lift. The only problem I found was that the foot was slightly unstable, and I was never able to get it to reset as it said it would in the patent.

Chapman hop pole puller with original cast-iron puller, Clayville.

# Harvest & Processing

One of the most rugged pole pullers is the "V-shaped" puller. These have been found all over the hop counties and vary greatly in size and workmanship. Most of these pullers were made from a simple iron bar or an old file. The metal part is shaped like a "V," usually with some jagged teeth on the inside of the V. The V is attached to a heavy wooden handle that is usually from 5 to 9 feet in length.

What makes this tool so effective is that the V is placed about 20 inches or so from the end of the handle. By placing the open mouth of the V on the pole and the end of the handle securely placed on the ground, a great amount of leverage is created, and the pole-pulling process is greatly facilitated.

An interesting variation of this tool is in the placement of the V. On most of these tools the open part of the V faces upward, but on some the open part of the V faces the ground. I do not know why this variation was done, but both types work well.

Long before I had a V-shaped puller in my collection, I saw one in a picture of Jimmy Clark's hop yards, south of Cooperstown. From time to time they are visible in other old images. Several years ago, on a farm near Cazenovia, Dave Petri and I found, in the corner of the kiln, fifteen large V-shaped pullers neatly stacked. It was almost like they were left there after the season was over and were ready to go for the next harvest. This was interesting because it showed how many of these tools might have been on a regular hop farm.

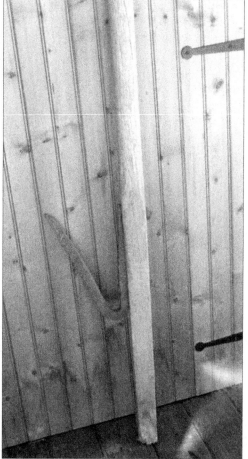

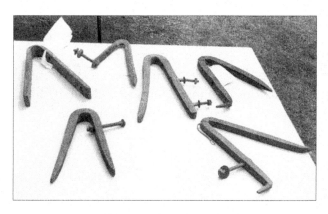

Examples of blacksmith-made V-shaped pullers.

V-shaped puller, Cody farm, Cazenovia.

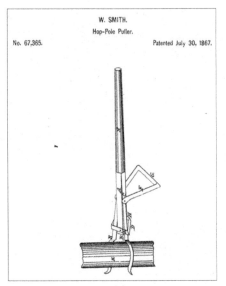

Patent drawing for a Smith pole puller described and pictured at right.

The last type of puller had a jaw on the end of a wooden handle and a metal support that hinged down from the handle to make a fulcrum. The metal jaw was placed at the base of the pole and the backrest was lowered into place. One needed only to push down on the handle to loosen and lift the hop pole from the ground. I have only seen a few of these tools, but never any marked examples. Apparently this idea was popular with inventors, since several patents for variations of this theme were granted. They all follow the same idea, but use different jaws and lever systems. The examples I have seen all appear to be early 20th-century in manufacture.

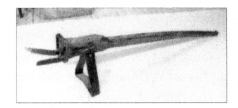

One type of puller that I have never seen, but which looks very interesting, was patented in 1869 by O.B. Hale from Malone. It is somewhat reminiscent of a hop dog, but much larger. It also has large open handholds, which a hop dog never has. The patent illustration is also interesting because of the type of shoulder strap that it illustrates. I mentioned this type of strap earlier under hop dogs. This tool never came into use in the Central New York hop-growing counties, but might have been known in the northern part of the state.

Patent drawing for a Hale hop pole puller.

# Hop Jacks

A tool that was often used with and near the hop box was the "hop jack." This was a device that held the hop pole while the box tender removed the hops from the pole, or to hold the pole while it was being picked.

The most common type of jack was a large-sized sawhorse. These sawhorses were about 6 feet tall and looked exactly like their namesake. The poles rested on the top of the jack and the hops could be easily cut off. This sawhorse jack appears in many old images, but very few exist today. One of the few extant examples is from the Wedderspoon farm in Pierstown.

A second type of hop jack was an all-metal variety. They were made of iron bars with a point on one end. The other end would have a large open "U" shape that would hold the hop pole. They were usually about 5 or 6 feet tall. I have seen them in old images but never had anyone tell me of their use.

Sawhorse hop jack on the Wood farm, Westville.

Unknown man holding a metal hop jack, near Cooperstown.

Blacksmith-made hop jack or pole support, Springfield, Otsego County.

The final type of hop jack I like to call a "pole rest." These were made from a sturdy hop pole. At the large butt end, two or three legs would be attached. Running horizontal to the ground, a piece was attached to hold the hop poles. The one example I have of this type came from a farm in the Town of Seward in Schoharie County. I also have collected a good postcard view of a hop rest in use from the Town of Maryland in Otsego County.

Hop pole rest in the center of picture, Palmatier farm, Maryland.

Hop pole rest, Seward, Schoharie County.

## The Sacker

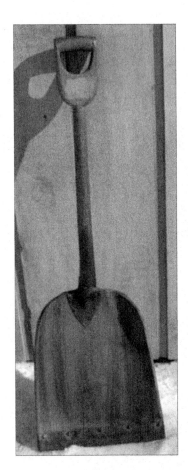

When the hop box was filled, the picker would call out, "Sacker! Sacker!" The sacker was the man that had the responsibility of removing the hops from the box and placing them into a hop sack. The hop sacks were homemade from hop cloth, which was a loosely woven burlap. On some examples the initials of the owner is stenciled on the sack. Each sack held 8 bushels of hops—the contents of one box. When the sacker came to the box, he would scoop the hops up in his arms and place them into the sack. The pickers and the box tender might help with holding the sack open to receive the hops. If any tool was used to take the hops from the box, it would usually be a large, flat-ended shovel. These were not special hop shovels, but the common coal shovel. The sacker also had the responsibility of giving the picker a hop ticket, or punching the picker's ticket, or recording the picker's box. This was important, since it was the basis for paying the pickers.

The hops, once sacked, were placed in a location that would not be in the direct sun, as that would make them "sweet" and cause damage. At the end of the day, or any time that a wagon-load of sacks was ready, they would be removed to the hop house for drying.

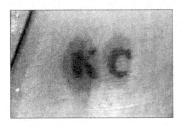

Stenciled hop sack, Keith Cody farm, Cazenovia.

*Harvest & Processing*

Robinson family, Exeter, Otsego County. The hop shovel was made by R. Rowland, who was in business from 1893-1901. It is the same shovel pictured on the facing page.

Hop sacks on the Buyea farm, Smithfield, Madison County.

# Hop Stoves

The drying of the hops was the most important process to ensure a top quality product. For many years in the 19th century, the quality of New York hops was based on the superior job our farmers did in drying the crop. In the hop house a variety of special tools and implements were used.

The hops would be brought into the kiln at the end of each day's picking. During the harvest a temporary platform was made along the side of the kiln that had the door into the drying floor. The sacks were brought up onto the platform and then taken into the drying floor. The hops would be dumped from the sacks and distributed in an even manner around the floor. The depth of each "charge" on the floor would be from 12 to 20 inches. No set depth was required, and in the old kilns you sometimes see a line on the wall of the drying area that indicates the depth that the hops should be piled. I have never heard of any specific tool being used to spread the hops. It was easily done by shuffling one's feet, or with a common, wooden barn rake. As long as the hops were not crushed or broken, it did not matter how this was done.

Because the most important job during the hop harvest was drying the hops, on many farms the owner took on this critical responsibility. In the furnace room the most conspicuous implement would have been the hop stove or furnace. These two terms were used interchangeably in the hop country and refer to the same thing. The stove was usually placed on the wall, away from the stovepipe hole and centered along the wall. Coming out of the stove would be a stovepipe, 3 feet long, which was connected to a series of stovepipes that formed a circle around the room, and was vented out the chimney hole. The pieces of stovepipe were held in place by heavy-gauge wire that was attached to the slats of the drying floor above. Hop house stovepipe material was very heavy compared to modern stovepipe. Many times the joints on the stovepipe were riveted for more strength.

The earliest hop stoves were the box stoves. They were constructed of eight cast-iron pieces. The base was a very heavy casting, about 2 inches thick, with two removable grates on the side that faced into the firebox. The sides were made of two iron plates, and a similar iron plate was used for the rear of the stove. The front panel was a cast frame that held a hinged door. The top of the stove was a single casting, with a 12-inch circular lip in the middle of the top, and two 7-inch stovepipe holes at the rear of the casting. The stove was 44 inches long and 19 inches wide, and stood 22 inches in height.

The only two examples known are held together with three large bolts that extend from the top plate to the bottom plate. In both of these stoves the side and end panels have raised diamonds in the panels. The door is typical of a hinged stove door. Neither stove has any type of air control on the stove. The flow of air was regulated from underneath the stove. In an example near Poolville, the original base setup is still present. A brick base is made with a iron door at one end that permits air to come up through the grates of the stove. These stoves are not marked in any way.

These stoves have grates inside to let ashes fall down into the base, and two stovepipe holes at the end away from the door. A sure sign that you have a hop stove is to look at the stovepipe holes; if you see two holes it is a hop stove. However not all hop stoves have two holes, some have only one, and a T-joint will be added to the stovepipe above the stove.

I had an interesting experience collecting the first box stove I ever saw. I was shown a picture of this stove in the early 1970s, but never went to look at it. Nearly twenty years later, the hop house with this stove was pulled down, since it was felt to be unsafe. The neighbors purchased the contents of the barn on the property, which included some hop boxes, and I went over and bought them. They said that the boxes had been in the old kiln. I asked about the stove that I had been told was in the kiln. They did not know anything about it. I went to the present owners and asked if I might look around the wreck of the hop house. With their permission the search was on. I remembered that the stove was in the middle of the building when last seen, so a focus point was established. The work of removing debris lasted most of the afternoon. Finally, late in the day, the stove was uncovered. The next problem was how to get it out of the mess of wreckage. A path had to be cleared and a moving cart used to move it. These stoves are very heavy and, with the help of only one other person, the task was very difficult. All this effort was worthwhile since, at that time, no other box stoves were known.

Only two complete early box stoves are known. A later box stove, not of the same manufacture, is in the collection of The Farmers' Museum. It is much finer in its casting, and the iron is of a better quality. Also this stove does not have the diamonds on the side and end panels. The door of the stove is marked "Waterville, Leonardsville N.Y." This stove

Box-type hop stove from the Teachout hop house, Westville, Otsego County.

was probably made by H.D. Babcock, the prominent foundry owner in those two towns. The stove also differs from the earlier models, in that it has a rectangular rim on its top, rather than a circular rim as the earlier stoves have.

Also at The Farmers' Museum are two examples of the second type of hop stove found in the Upstate area. The two stoves are both from Otsego County and are similar in design, though from different makers.

The bulging fire pot in the center of the stove characterizes the Otsego type. In some cases, the stoves are also marked with the words "The Otsego," or just "Otsego," and a number on the base or ash pit. These stoves come apart into several pieces. They are heavy enough that the weight of the parts alone will keep the stove in place. They have a base of about 16 inches in height, with a door in the front for ash removal and air regulation. In the door is also found a shaker to move the grates inside the stove. These grates are very heavy, weighing 75 pounds. On top of the base rests the first of two rings that make up the fire pot. The diameter of the rings varies, but the lower ring is always the heaviest, often almost 2 inches thick. The upper ring rests in a lip in the lower ring. Both rings are beveled. The lower is smaller at the bottom and wider at the top, with the top ring reversing the bevel. On top of the fire pot is a cast-iron section that supports the front door of the stove used to feed the fire. (On those stoves that have maker's names, they are found on this door.) The final piece is the top, which has a very distinct appearance. It is slightly convex, with a deep flange around the outer edge. Also in the top are found two stovepipe holes. In some stoves only one large stovepipe hole is found. Stoves with a single large hole were made in Waterville and Oriskany Falls.

The two examples in the museum collection represent two competing manufacturers, but are of the same style. The larger of the two stoves was made by Risley & Company, Richfield Springs. It is an Otsego No. 21. The stove might burn either wood or coal. In a pamphlet by the manufacturer from the 1880s, it gives the price of the stove as $40.00. It is interesting to read the testimonials in the pamphlet, as they tell us of the improvement this type of stove was over the box style. One contributor noted that the Risley stove saved him nearly one-half the fuel he used in the box stove. Other points they stressed were that the time in attending the stove was reduced, and that much less stovepipe was needed to get the heat distributed.

Otsego-style hop stove, Risley & Co., Richfield Springs.

Advertisement for an Otsego-style hop stove, showing interior grates and a jar for holding sulfur sticks at bottom left.

Otsego No. 21, Risley & Co.

Door from an Otsego No. 20 hop stove, Fly Creek.

The second example is an Otsego No. 20, made by Niles & Babcock of Fly Creek. This is a slightly smaller stove. It is about 46 inches high, while the Risley stove is 53 inches in height. In construction and style it is very much like the Richfield stove.

Other manufacturers are also known. In Leonardsville, the firm of H.D. Babcock was a well-known maker of hop stoves and other equipment. In my travels in the hop region, I have seen more Babcock stoves than of any other manufacturer. They had their own name for the stove rather than The Otsego, they called their stove "The Granger," but the style is still the same. The Granger came with a one- or two-stovepipe setup. Some are only marked with "Leonardsville," others say "Waterville, Leonardsville, N.Y." On an example in the museum collection, the bottom door is marked "Coal Hop Stove No. 1." On the frame of the door it reads, "Pat. Apld. For."

In Schoharie County, stoves were made in Cobleskill and Richmondville. The Cobleskill stove was marked "Manfd by Ira D. Rickard Cobleskill NY" on the upper door. The basic design was the Otsego style, with two stovepipes. The only parts of the Cobleskill stove that differed from the Otsego stove were the base and the top door. On the Cobleskill stove, the sides of the base angled out and were squared off, creating a larger base. In other Otsego stoves, the base was rectangular in the front, but round on the sides and the rear of the stove. The vents on the doors of the Cobleskill stove were three in number, and rectangular. In other Otsego stoves, the vent was circular, with three small triangular vents and a turning control. I have seen stoves with no name but with all the characteristics of the Cobleskill stove, most notably the base. I have only seen one Richmondville stove, but did not get the manufacturer. It was not significantly different from the other Otsego stoves.

Cobleskill stove. This one was not marked, but the base is characteristic of that type of stove.

Advertisement for a Cobleskill hop stove.

Another important manufacturer was in Munnsville. The striking characteristics that set these stoves off from the Otsego type were the use of a ridged casting rather than a smooth surface on the upper section, and no bulging fire box. The stove was also held together by the use of long, threaded iron bolts. Also, in the midsection, they were more elongated than the Otsego style. The Munnsville stove was made to burn coal, which was heralded as a great improvement. In 1873, Stringer, Barr & Company developed this style with coal-burning grates. Two years later they patented this idea and, for many years, did a good business under the name of the Munnsville Plow Company. Later, the firm was owned by Stringer, Dexter & Company, but kept the same firm name. These stoves, for all their popularity in the 19th century, are very rare today. I have seen only three examples of Munnsville stoves, and they were very different in size.

Munnsville hop stove.

Another interesting stove was "Kelley's Improved Radiating Furnace," made in Oriskany Falls. Three features made these stoves unique. One was the way that the fire pot was made, with horizontal lines on it. In the Otsego models, the fire pot is always smooth-cast. The use of a track that permitted the door to the upper part to move around the body of the stove also differed from the Otsego type. In the Otsego stoves, the door was hinged to the upper part. The final feature was the decorative cast eagle found on the upper section of the stove. I have only seen one stove like this, but have been told of these eagles by several people. Also on the Oriskany stove was a very interesting cast-iron T-joint stovepipe holder. The Oriskany stove usually had only one large stovepipe hole.

Oriskany Falls hop stove with a cast-iron T-joint stovepipe support.

During the drying of the hops the stoves were heated to a very hot temperature. The inside of the kiln reached a temperature of 140 degrees. People tell of the cherry-red color of a hot hop stove. Since a high temperature was required to dry the hops, it is not hard to envision such a hot stove.

The ridge on the top section of the stove was an important feature of the stove. During the process of drying the hops, sticks of sulfur or brimstone would be burnt. The top of the stove was designed as a place to burn the sulfur. This was not usually done, as the stove became too hot and the sulfur would burn too fast. Sulfur was burned for a variety of reasons. It bleached the hop and produced a uniform golden color. The brewers felt that this was a sign of a quality hop. It also made the drying process faster, and helped kill any bugs that might have been in the hops.

The lip on the top of the stove was for holding a "sulfur pan" or "jar." This was a slightly curved, iron container that had a hook on its back to hold it on the stove. Sulfur was placed in these jars, and a red-hot poker or piece of iron was placed on the sulfur to ignite it. With the purchase of each Risley stove one jar was included, and others might be purchased for $2.00 each. Other jars were also used. A wedge-shaped jar that fit on the top of the stove and rested on the lip was also used. It is also possible that round sulfur burning pans were used on the box type stoves. The raised circle on the top of the stove might have been a place to put such a pan.

If one did not want to use this method of burning sulfur, a very common system was to have cast-iron kettles hanging from the slats of the drying floor by a wire hook. A kettle was placed in each corner of the kiln at a height so no one would hit it. The kettles were taken down to start the sulfur, and then placed on the hooks. To

Sulfur pan or jar.

properly sulfur a floor of hops, 1 pound of sulfur would be used for a drying, more if the hops were rusty. In many an old hop house, the presence of wires still hanging from the drying floor above reflects the popularity of this practice.

One other feature of the furnace room that should be noted is the heat shield that is sometimes seen in the old kilns. This was placed directly above the stove or furnace. It is usually a very simple piece of sheet tin, suspended from the slats above by wire. The purpose of the shield was to protect the hops directly above the stove from the excessive heat generated by the stove. It also protected the hot stove from dust from the drying floor that might get on the stove and burn, producing an odor that might get into the hops above. The shields are of interest, since they clearly tell where the stove was placed in the kiln.

Wedge-shaped sulfur jar.

The "floor" or "charge" of hops on the drying floor was from 12 to 18 inches in depth. The hops would be loosely and evenly distributed around the floor. It would take twelve hours to complete the drying process. The purpose of drying was to reduce the water content of the hop. Green hops—those freshly picked—contain 70%-75% water. By drying, the moisture content was reduced to 7%-10%.

The heat had to be built up slowly in the first few hours of drying. During this time the ventilators on the top of the kiln had to be open to let the steam from the hops exit freely. At the same time, the sulfur had to be burning. When the hops stopped steaming, the ventilators were closed halfway, and the sulfur stopped. This would go for about ten hours. In the final two hours, the ventilators were totally closed to ensure that the top layers of the hops were dry. The heat from the furnace would also be reduced. When this process was completed, the vents were opened. After cooling on the floor for an hour, the hops were removed to the storage room.

After the hops stopped steaming, it was a common practice to "plow" the hops. In 1966, 91-year-old Lyman Wiltsie demonstrated to me his technique.

> You started around the edge of the drying floor with both feet on the cloth covering. This was important, as you did not want to damage the hops by stepping on

them. You then started to shuffle your feet through the hops. Be careful not to lift your feet from the cloth. This was done all around the kiln, making smaller squares each time. As you plow, the hops are stirred up and the top and bottom layers mixed.

It was hoped that this would ensure an even drying.

## Hop Rakes, Shovels and Pushers

In some cases special tools were used to help turn the hops. Rakes were probably the most common tools used. The everyday wooden barn rake was also often used. The rake was also helpful in leveling the hops on the drying floor and removing them from the floor. If a rake was used in a kiln, it is easy to tell today because of the black resin that the hops left on the tines.

A special rake was also used in the hop houses. This rake was made with a regular fixed rake and a movable second rake. Both rakes were 26 inches long. The non-moving rake had twenty-six teeth, while the moving rake had twenty-five teeth. In both cases the teeth were $3^{3/8}$ inches long. Both rakes were made of hickory. On the bottom of the handle was a special cast-iron holder that braced the fixed rake and provided two open, circular holders for the second rake, 3 inches up the handle. The handle measured 66 inches in length. Exactly how this rake was used is not clear, but the use of these rakes in hop kilns is obvious.

Top of a hop rake with a label of "The G. A. Swineford & Co. Canton O."

The first time Dave Petri and I found such a rake was on the drying floor of a hop house in the Town of Hartwick. Both a double rake and a wooden hop shovel were found together on the drying floor. That rake did not have the second set of rakes, but soon a second example was discovered on the wall as a decoration in a shop in Cooperstown. It was found that this particular rake came from a hop house in Springfield. Since that time, several other examples have been found. On sev-

# Harvest & Processing

eral of the rakes the dark resin left by the hops was clearly evident.

Another tool that was used to turn the hops was the hop fork. I do not have one of these and have only seen a picture of the tool, but have been told about them by several informants. The hop fork had a series of long tines, about 30 inches or more in length. They were all connected to a piece that held them in place, about 5 inches apart. A long handle extended from the tine support. These forks were pushed into the floor of hops, and the hops lifted and mixed with little damage. The only example I have seen was from Schoharie County, but I have been told about them in other counties.

Another style of rake was the type with an offset handle. The example pictured below came from Sangerfield and was used in a large kiln. It has seven teeth, 6 inches long. Six are round and have curved, pointed ends. The seventh is made from the end of a brace and is flat and pointed. The handle is placed at an angle of 45 degrees from the rake, and runs toward the back of the tool. It is 53 inches long. Halfway up the handle is the brace that extends down to the rake. At the end of the handle is a handhold at right angles to the handle. The rake itself is 42 inches in length. Some resin still shows on the teeth of the rake. I was told that a walk ran in the kiln above the drying floor, and this tool was used by a man on the walk to turn the hops. As far as I know no other examples of this tool are known.

Hop fork, Cobleskill area, Schoharie County.

Large offset hop rake, Sangerfield, Oneida County.

Remnant of a small hop shovel. The back is 15 in., the side is 19 in., both 6 in. high.

Another important tool used in turning the hops was the hop shovel. The shovel came in two basic styles, but there was a huge variety in the way they were made.

The shovels that were used to turn the hops were the small wooden types. Over the years I have found several examples of this style of shovel, and in some 19th-century literature I found descriptions of their use. I had always thought that they were to get into the corners of the drying floor, but that would only have been a secondary use. These shovels are about 13 inches by 13 inches square, with a rather short handle of 25 inches. Some have a handhold at the end of the longer handle. The sides and the rear of the shovel have a gallery of 5 inches, rounded at the front of the shovel.

In addition to the small wooden shovel, larger wooden shovels were also used. The larger shovels were mostly used to move the hops from the drying floor into the storage areas. Some are basically rectangular in shape, with a long handle and a gallery of about 5 inches. Others are longer and thinner in shape. In all cases, the wood that was used is fairly thin, to keep the weight down. It would seem that fewer of these wooden shovels were used, as they are not very common today.

The most common style was the frame shovel, with some type of a cloth bag or covering material to catch the hops. The size of these shovels varied greatly. Some, with their handles, were more than 6 feet long, others were much shorter. This classic hop shovel had a simple frame, often a yard or more in length. The corners of the frame were reinforced with tin straps, and a cross-piece near the rear of the shovel was used as a handhold. From the back of the frame extended a long handle, usually 3 feet long or more. Tacked to the frame was a large bag made of burlap or bed ticking material. The capacity of this type of a shovel would be quite large.

Basswood hop shovel, Otsego County. The handle is 54 in. long, the shovel is 27 in. by 20 in.

Hops are a very light, but bulky crop to move, so a shovel that might make a large scoop would be very handy.

Hop shovel with original cloth bag, Otsego County. Shovel is 29 in. by 37 in.

Left: Man demonstrating the use of a hop shovel shown in the photo above.
Right: Cloth-lined hop shovel. Both images c. 1900; from the collection of Robert Seaver.

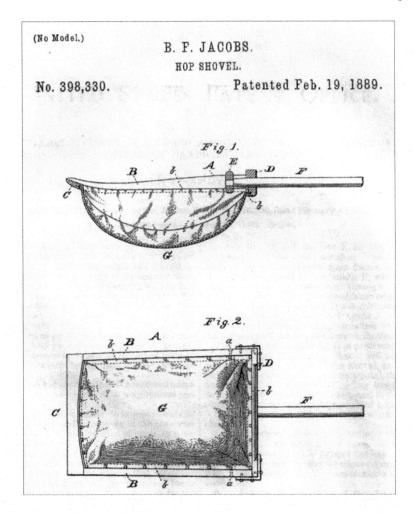

Hop shovel patented by Benjamin F. Jacobs of Milford, Otsego County.

Many variations of the shovel pictured above still exist. I have seen shovels that are dovetailed at the rear corners. Some have mortised joints. Some are curved, rather than being straight on the sides. In others, the frame to support the cloth covering is very high and elaborate. Some very interesting shovels are made from other farm tools, usually an old barley fork. By putting cloth under the tines, a shovel was made. In one case, extra tines were added to make a larger shovel. As far as is known, these were made at home, and the ingenuity of the creator was the only limiting factor.

(Facing page.) Top, right and left: Hop shovels on display at The Farmers' Museum Hop Seminar (1995). Bottom, left: Illustration of a man with a hop shovel, from *Harper's Weekly*. Bottom, right: Hop shovel from Richfield Springs, Otsego County, similar in style to the one in the illustration at left.

# Harvest & Processing

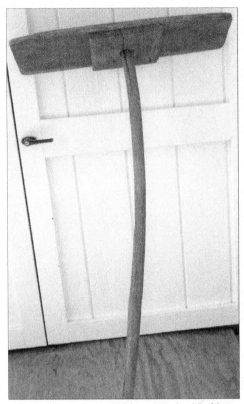

Hop pusher, from Bridgewater in Herkimer County. The handle is 62 in. long, the pusher is 32 in. by 9 in.

The last tool commonly associated with the drying of the hops is what I call a "pusher." The purpose of this tool was to push the hops from the drying floor into the storage area. I would say that all the pushers I have seen are homemade and vary greatly in style. Basically the pusher is a long, flat board attached to a handle. Often a metal support was attached from the handle to the pusher to make it more stable. Usually the handle was quite long, more than 50 inches. Sometimes an old barley fork handle was used, with the pusher attached by the shortened tines.

The floor of the kiln was built higher than the storage room. Usually you find one or two doors that lead from the kiln into the storage room. No standard height existed for the difference between the two floors. In some cases, the drying floor is almost 6 feet above the storage area. In some of the older hop houses, the drying floor is hardly raised from the storage floor.

In a few hop houses you will find hinged sections of the kiln wall. These are level with the drying floor. During the drying they would be closed down, but once the drying was complete, they would be opened and the hops easily pushed from the drying floor into storage. A large hop shovel or a pusher would make quick work of a floor of hops if the kiln had this type of an arrangement.

After the hops cooled for an hour they were ready to move into storage. Care had to be taken not to break the hops when moving them. On the storage floor they had to be spread out evenly across the room. As the drying progressed, the hops were layered into the storage room. The hop rakes, shovels and pushers might all have been used for this job.

The harvest and drying lasted for several weeks, and the hops were picked at differing states of ripeness. The earliest hops were less ripe than those picked later and had less value, so mixing the floor charges created a better grade of hops overall.

The harvest and drying work was now done, and the grower could look toward the tasks of the fall season.

# October to December: Baling and Selling

With the end of the harvesting and drying of the hops, the farmer's pace might slow down. Usually the hops were not baled until sometime in October. The month of September, after the drying was complete, was not a particularly busy time for the hop farmer. In Leon Van Patten's diary he only mentions two specific hop-related chores once the crop was dried.

Leon had finished picking on September 7. On September 22 he made the following entry in his diary: "Manured the hops and stacked the poles." A hop plant required a large amount of fertilizing during the year to reach top production. When the bines were cut during harvesting, they had only about 10 inches or less left above the ground. As preparation for the winter—and as a source of fertilizer—two forkfuls of well-rotted manure would be placed on each hill. This would protect the plant during the cold months of winter. It would also rot down, leach into the ground, and become food for the plant.

The poles were stacked as a way to preserve them and hopefully get more years of use out of each pole. No special tools were used in stacking the poles, but the methods of stacking varied. Some stacks were very solid and, when viewed, did not let any light through. Others remind one of an Indian teepee. At any rate, the well-stacked poles marked the change of the season from summer to fall.

View of hop pickers showing poles stacked for the winter season.

The second entry in Leon's diary on September 26 reads, "Burned hop vines." The bines, as they were cut from the poles or the wires, were piled to the side. In many of the old pictures the piles of hop vines are very noticeable. After the bines had dried for several weeks they had to be destroyed. These bines could be a source of disease, mold, or insects. They had no value as animal food and little as fertilizer, so the easy way to destroy them was to burn them.

In September, other work related to preparing the farm for the winter had to be started, but the hop-related jobs were limited. As the cool, damp days of October approached, another critical task awaited the hop farmer: baling the crop.

On October 12 Leon records in his diary that he started to bale the hops. The work progressed until October 25. At that point he had baled 43 bales and felt that only 2 bales more were left to finish the job.

This pattern of waiting a month or more before baling the hops was a common practice among the small farmers of the Upstate area. After the cool hops were removed from the drying floor, they had to "sweat" in the storage area. This usually would take about ten days, depending on weather conditions. The hops would toughen and take on some moisture as they cured in storage. At any point after this they could be baled. In the hop region, the common practice was to wait until October, rather than to rush to bale.

## Baling the Hops

The earliest method of preparing hops for market followed the European model. Bags called "pockets" were the common way to send hops to market in Europe and was the method used at first in the United States. From 19th-century accounts it is well known how these pockets were filled.

A circular hole was cut in the floor of the storage area of the hop house. The pocket was placed in that hole with the open end secured around the edge of the hole. The hops were put into the open pocket. To press them down a man would climb into the pocket and tread around with his heels striking the bag. This would be done as needed until the pocket was filled.

This practice ended early in the 19th century and no artifacts exist from this process. The closest link to this early system that I have ever found was in a hop house in Montgomery County. It had a round hole to let the hops down from the storage area. It is the only time that I have ever encountered a round hole in all of my investigations of hop houses. The hole was not large enough for a man to go into, so I am sure that it was not used for pockets.

## Screw Presses

The first major development in baling technology was the introduction of the screw press. These were very common by the third decade of the 19th century, and did not go out of use until well into the 20th century. There were two types of screw press, portable and stationary.

The search for the portable screw press is an ongoing quest. In all my travels over the hop region, I have never seen a portable screw press. I have been told of them by many people, but have never found this artifact. The descriptions of these presses vary. In some accounts they have two wooden screws, in others only one. The top piece of a press was collected in Brockman's Corners, but it is not an absolute that it was part of a hop press. The strongest evidence that it was part of a hop press is its size. It is long enough to be the size of a hop bale.

Top piece of a press from Brockman's Corners.

The idea of a portable screw press must have been quite popular, and one widely respected in the hop-growing areas. In the February 26, 1859 edition of *Moore's Rural New-Yorker*, such a press is pictured (see following page). Lincoln L. Cummings of Munnsville patented the press on June 15, 1858. The article gives a detailed description of the operation of the press. One of its best features is that the press does not get jammed when the top is not pressing evenly. The press was well received at county fairs in Otsego, Madison and Oneida Counties. An advertisement for the press stresses that one man may take it apart and load it into a wagon. With two men working, 20 bales of hops could be finished in ten hours. The cost of the press was $40.00. To date no examples of this press have been found.

Illustration of a Cummings portable hop and hay press.

Stationary screw presses were also being used at the same time as the portable screw presses. In many of the early-style hop houses, the presence of a large beam, usually across the center of the pressing room, with a threaded hole in its middle, is a sure sign that a screw press was once in that building. A second indicator of the presence of a screw press is a square notch on the beam, and a round hole that extends through the beam into the storage area above. In a few rare cases the notch will contain a cast-iron piece that has a threaded hole in its middle. Today, the largest number of buildings that show evidence of screw presses are in Otsego County and the Town of Minden in Montgomery County. At one time screw presses would have been abundant in the entire hop-growing area.

A press beam is not constructed like the normal beams in a hand-hewn building. First, they are much heavier in construction than the usual carrying beam. Many times these beams are as much as 20 inches or more on all four sides. They are also often thicker in the middle than on the ends where they are joined into the building. I would assume the need for extra strength to support the screw was the reason for the thicker middle.

From examining these beams it has become clear how the screw-holes were made. The hole was first roughed out by drilling a series of smaller holes around the circumference of the hole's outline. This was probably done with a 2-inch hand drill. On a few beams today the marks of these drilled holes are easily seen. Once the hole was opened, the next step was to make the thread in the hole. I have never seen the tool used to make the threads, nor have I ever been told by anyone how it was done. I would imagine that a tool like the "Cider Press Screw Tap" pictured in Mercer's <u>Ancient Carpenters' Tools</u>* might have been used to make the thread. This tool had a frame to hold the round tool in place as it was screwed into the hole. Attached to the tool was a cutter that would make the threaded groove as it was forced down into the hole. The device looks much like a hop screw without the threading. In Mercer's book he identifies the tool as being used to make a threaded beam for a cider press. This would be very similar to the beam needed for a hop press.

Into the threaded hole a long, threaded beam would be screwed. The bottom (or head) of this beam was several inches larger than the threaded part. It also had one or two rectangular holes through the beam. In most cases, there were two hand-forged iron bands, one above and one below the holes. These bands were to strengthen the beam so that it would not split.

The diameter of these beams varies and the construction indicates that several shops produced them. Some have very long heads, others are much shorter, and one has a tapered head. At the other end that goes into the storage room, most screw beams are simply cut off. In the area around Salt Springville, in Otsego and Montgomery Counties, several screws have been found with unthreaded tops of about 17 inches in length. These screws are believed to have been made in the shop of W.S. Weller of that hamlet. Four of these screws have been found within just a few miles around the Salt Springville area.

Two methods of threading are evident in the screws. Some appear to have been turned. In these beams the tool marks clearly show the use of turning tools. On other examples, the threads have been cut out or at least finished with chisels. This is easily noticeable from the marks left by the slightly-curved chisel blade. So far no marked examples are known. The geographic distribution of the screws is also interesting. I have collected a screw from near Duanesburg in Schenectady County in the east, to as far west as Pompey Center in Onondaga County. Most of the known examples are from the area of northern Otsego County, into Montgomery County.

---

*Mercer, Henry C. *Ancient Carpenter's Tools*. Doylestown, Pa.: Bucks County Historical Society, 1960, p.232.

Wooden hop screw, Van Deusenville, Otsego County. Right: Wooden beam showing the threaded screw-hole for the hop screw pictured above.

The second type of screw is a cast-iron screw with a wooden head. As I noted above, several examples of screw-holes that were made for iron screws are known, but only one iron screw has been found so far. It was found in a wonderful old hop house just outside the hamlet of Salt Springville in Otsego County. The beam in the press room still has the iron-threaded nut in the press beam. The screw itself was buried in ashes in the furnace room of the building. It was covered so that only the wooden bottom showed. It is threaded, with a round ball at the bottom of the threads. In the ball is a hole to place the handle to turn the screw. The bottom of the screw is wood, with two iron bands for strength. As with many of the wooden screws, a small projection on the bottom is present to hold the screw in place during pressing.

Iron hop screw, near Salt Springville, NY. This is the mate to the cast-iron threaded nut pictured above right.

Placed on the floor of the press room, below the screw beam, would be the press box. Only two of these early-style press boxes are known. In each case they were basically built the same, with differences only in minor details.

The press box had to completely break down so that the bale could be sewn together while the pressure was on the compressed hops. The floor of the press was made of boards 2 inches thick, and rested on two cross-pieces that extended out beyond the floor planks. These cross-pieces were notched to hold two upright pieces. These uprights had a tenon on their top that fit into a mortise on a cross-piece that held the two uprights securely at the top. In between the two uprights was an end panel that was 17 inches by $51^{1/2}$ inches, and the butt ends of the side boards. On an example at The Farmers' Museum, the uprights are connected along the sides by a board 4 inches wide. On another known example, this connecting board is not used.

When assembling the press box, the uprights were placed in the base, then the sides would be built up and the end plates put in. After this was complete, the top cross-piece would be put on, and the press box was ready to use. Before the box was put together, hop sacking was laid across the floor, with enough cloth overhanging the press to come up the side of the bale more than halfway.

The screw press was a slow way to bale hops. From the hole in the storage floor, a cloth chute would hang. This would direct the hops into the press box. One man would be in the box as the hops were being pushed down the chute from the storage area above. This man was known as a "jumper." His job was to level the hops in the box and make sure that the corners of the box were filled. With his feet he had to tamp down the hops, being careful not to break the fragile cones. Sometimes a board would be used in the press to help the process of compressing the hops. When the press box was nearly filled, the hops could be pressed by the screw.

A covering known as the "fowler" was put into the press. This fit the box very snugly. On top of the fowler, additional planking was built up so that the screw did not have to be brought so far down. The screw was then turned down with the help of a long turning-handle that fit into the notches in the head of the screw beam. This first pressing usually did not fill the press box, and more hops would be added until the required weight was achieved. In the final pressing, the top piece of bale cloth was placed over the hops before the fowler was put in. Then the screw was pressed down again, and the compressed hops were held in place by the pressure of the screw beam.

The cross-piece at the top of the ends was then removed and the uprights were loosened. It was now possible to take out the sidepieces and let the bale cloth fall down the side of the compressed bale. At this point the bottom piece of cloth was sewn to the top piece on both sides of the bale.

Early style press box.

The screw beam was then released, and the hops would push up and be held tightly by the bale cloth. The ends of the press could then be removed. The bale cloth, slightly longer than the side of the bale, was now folded and turned over the ends of the bale. To secure it, pins were driven into the end of the bale. For each end, a piece of bale cloth was cut to a size that would fit over the end, and then sewn into place. The bale was now complete.

The process with the screw beam was rather time-consuming. I have never seen any statement as to how long this operation took.

Above, left: Dave Petri placing a wooden screw into the beam at a hop house in Van Duesenville. Above, right: Installed hop screw and press box, Van Duesenville. Bottom, right: Hop screw found near Salt Springville, possibly made in that village.

# Ratchet Presses

Shortly before the Civil War, a major change in hop pressing took place. In August of 1860, Lewis W. Harris of Waterville received a patent for a new ratchet-type press. This press had the advantages of being both portable and easier to operate than the screw press. The Harris press became a mainstay throughout the New York State hop-growing regions. It was made by several firms and would continue to evolve until the end of the century.

Harris hop press at The Farmers' Museum Hop Seminar (1995).

The basic Harris press had a bed-piece that was the base for both the floor of the press and the rest of the press. Two solid pieces made up the sides. They were attached to the base so that they could be raised or lowered during the pressing process. The ends were made up of a series of boards that could be added as the amount of hops in the press increased, or placed in at the beginning of the baling. On either side of the ends were two cast-iron ratchets attached to wooden uprights. These extended above the press sides. Sliding down over these uprights was the top piece that would push down the fowler and compress the hops. On either end of the press were attached two handles that had devices that lowered the top plate down, much like a car jack does.

All the parts of this press were very heavy. In many cases the top base and side uprights were made of maple. The sides were usually a soft wood, but very thick. These presses do come apart, but from experience I know it is quite a job to move one.

Across the bed-piece of the press a piece of hop sacking, 57 inches by 72 inches in size, was centered on the floor of the press. The sides of the sacking would be parallel with the sides of the bed-piece. Then the sides and ends were put in place, and the hops put into the press. Usually the press was filled about halfway at first. Then the jumpers got into the press on a snugly-fitting board, using their weight to compress the hops. The board reduced the problem of damaging the hops. This process was repeated a second time. After the second filling, a third filling went into the top of the press box. The fowler was put in place and the press was ratcheted down. Two men—one on either side—would operate the handles and lower the press piece in unison. The "click … click … click" sound of the handles ratcheting down the press was a familiar sound during baling.

The pressure was then released, again the hop press box was filled, and the top piece of hop sacking was placed over the hops. Then the second pressing took place. The hops were pressed to a point a little less than $2^{1/2}$ feet from the bottom. Finally, the sides and the ends of the press were removed. The hops were held in place by the pressure of the press.

Left: Illustration of a Harris-type hop press with the front open, from *Harper's Weekly*. The man kneeling is sewing the bale cloth together, to the left is the scale used to weigh the bales.

Harris-type press, open view.

## Other Presses

Manufacturers marked many hop presses with stencils.* The large, orange-colored press pictured on the following page was made by one of the most common manufacturers. This press is marked "B.A. Beardsley, Waterville, NY." This press is a very late example. The slip plates that close up the ends of the press box are metal plates that extend from the floor of the box to the top of the sides. These plates came in during the 1890s and are a good indication of the age of a press. Also the top of the sides have fixed ends that flair out at an angle to help direct the hops into the press box. Again this was done on the newer presses. In the earlier presses the ends are sealed by a series of boards, usually about 4 or 5 inches wide, which are built up to make the ends of the press. The angled pieces at the top of the ends are removable on the earlier presses. As with any invention over the years, some improvements were made and help show the evolution of the tool.

---

*Hop bale stencils are discussed in detail beginning on page 118.

Beardsley hop press made in Waterville.

A Utica firm made the second example of a stencil-marked press. The makers were Leroy, Shattuck & Head. This press is very notable for the bald eagle stencil and the date of 1876 on the side panel. Only two of these presses have been found, and both came from the same neighborhood of Pierstown in Otsego County. The color of this press is a dark, reddish-brown, to an almost deep red.

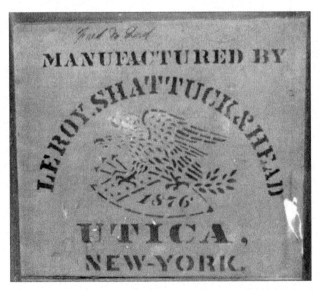

Stenciled image from the side of a hop press made in Utica.

Two Cobleskill press makers are known. In both cases the presses are of the Harris type. One maker was E. Waldron, the other was Ira D. Rickard. In many cases the manufacturers of hop equipment made a complete line of hop-related items, and the same name would appear on presses, stoves and other items. Both of the above presses were orange or deep red in color, but some presses were painted green, while others had no paint on them. Other types of hop presses were made in competition with the Harris press. These presses never enjoyed as wide a popularity as the Harris press, and are extremely rare today.

I had never found any non-Harris press until 2002. I was called to look at some hop screws in the Town of Minden. When I examined the screws, I was further intrigued by a pile of light-gray pieces that were with the screws. These turned out to be parts of a press made in Springfield Center, Otsego County, by the firm of Shipman & Son. Only the base, part of the press box, one cast-iron ratchet and a few other pieces were left of the press. When I went home I found that I had the patent papers for this press. It was patented on November 26, 1861. It was clear from the papers that more than half of the press was missing. This press worked on a very different concept than the Harris press. The floor of the Shipman press did the pressing, not a fowler as in the Harris press. The ratcheted sides lifted the floor of the press and compressed the hops. The sides would be removed and the bale sewn together before pressure was totally removed. The bale size was the same as the Harris press.

Stencil image from a Shipman hop press.

Another type of press was the McCabe press. James E. McCabe of Clinton patented this on March 11, 1879. The press box was much like the Harris press, but the method of pressing was different. In this press, the fowler was raised and lowered by turning a crank on the end of the press. This was geared into a large wheel that moved the fowler down a ratcheted bar inside the press. The fowler in this press was made with hinges so that it would fold out flat to press and drop down as it was

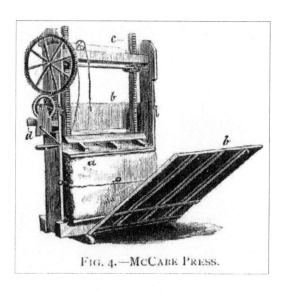

Fig. 4.—McCabe Press.

raised. This had an advantage over the Harris press in that no jumper was required to work in the press box. It was hoped this press would be faster in operation, but some debate on that point existed. In the 19th-century literature, the Harris press was considered to be a speedier piece of equipment. The McCabe press was also heavier than the Harris press, and cost about $30.00 more to purchase. I have been told about these presses but have never seen an example. B.A. Beardsley of Waterville made this style of press.

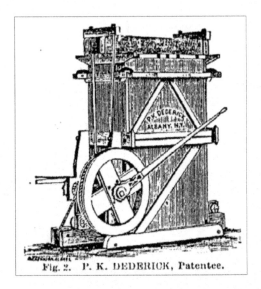

Fig. 2. P. K. DEDERICK, Patentee.

P.K. Dederick of Albany made yet another style of press. This press was sold for both hops and hay. It was called a "Progressive Lever Hand Press." The name reflects what made the press different. It was "progressive" because it built up the pressure as it compressed the bale. The idea was for the press to work rather quickly at first, and then build up power as the bale was compressed. The press had a hand-operated lever system that pushed the floor of the press up against a stationary fowler. Once the bale was pressed, the sides would be removed, and then the bale could be stitched and removed. There were problems with this press. Its weight, at 800 pounds, would have made it a very heavy piece of equipment. In addition, the price was $100.00—much more than the Harris press. No examples of the Dederick press are known.

## Bale Pins and Hop Needles

After pressing, it was time for the hop bale to be sewn together. The top and bottom pieces of hop sacking overlapped each other and were held in place by bale pins. These look something like meat skewers. They were usually 6 to 8 inches long, with a loop on the end. The other end was pointed and would easily go into a bale. Some pins had wooden handles and a metal point. In a few cases, very long pins were used.

The men who did the baling also sewed the bale. Long, slightly-curved needles were used. These are always referred to as "hop needles" in the hop region, but they are the same needles that were used to sew grain bags. Most of these needles are made of steel and many bear the mark "Made in England." I have seen hand-forged hop needles, but only a few. They come in a great variety of sizes, from 4 to 8 inches in length, with variations in their curve and the shape of their head.

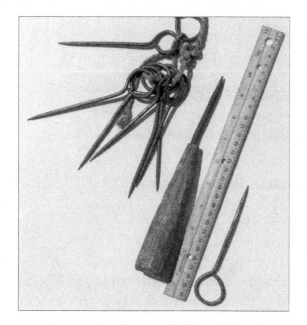

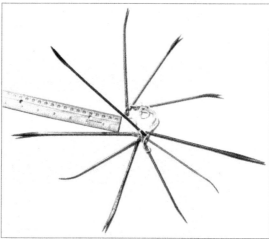

Left: Hop bale pins. Above: Hop needles of various sizes, made of steel and iron.

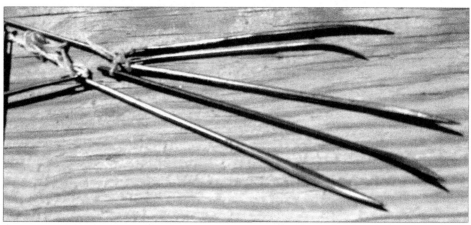

Close-up of hop needles.

A second type of hop needle was long and had a wooden handle. These were also known as "sail needles" in coastal areas. They had a bulbous wooden handle of $4^{1/2}$ inches in length. The needle was fastened into the handle and extended out for $11^{1/2}$ inches. The tip of the needle was flattened $2^{1/2}$ inches before its end, with a slight hollowing on each side. In the hollowed area was a hole. The point of the needle was 1$^{1/2}$ inches long and tapered on four sides to a very sharp point. The only mark found on examples of these needles is "Chatillon," near the handle. Apparently they used a different stitch than the lock stitch used with the curved hop needle. Since fewer of these large needles have appeared, it is obvious that they were not used as commonly as the curved hop needle.

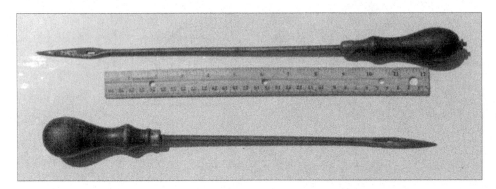

Hop needles, Otsego County. The top needle is marked "Chatillon."

Before releasing the press, the sacking had to be turned over the ends and lapped to make square corners. These would be held in place with wooden pegs driven into the bale. Another way to hold the ends fast was to take the twine at the end of the seam on the side, and extend it across the end to the seam on the other side. Two prominent growers, Morris Terry and C.L. Terry, developed this method in Waterville. In both methods, a separate piece of sacking was sewn over the ends to finish the process, after the bale was removed from the press.

Once the sewing was complete, the pressure of the press was relieved, the hop sacking was stretched tight by the expanding hops, and the bale took its correct size and shape. The bale was then removed from the press, and the ends sewn on.

## Hop Stencils

Hop stencils, used to mark presses and bales, were of two types. A solid brass stencil was often used. They varied in size and the information they held. The large brass stencil pictured on the facing page gives the name of the farm and the town, but some also included the county. Smaller brass stencils might only include the owner's name. Many times the name stencils were used on the walls, doors or stairs of the hop houses, and can still be seen today.

Brass hop stencil.

The second type was a stencil made from brass letters that snapped together to form the owner's name and any other information they wished to give. These letters were placed in a tin box that separated them until needed. This was an inexpensive way to make a stencil that had a lot of flexibility. On occasion, stencils were also kept with the sampling tools in a kit.* The stencil kit might also have a long bristled brush to apply the ink, which was made of turpentine and lamp-black. Generally the owner placed his name on the side of the bale.

Stencil kit.

Brass hop stencil, Smith & Capron, Oneonta.

---

*These hop sampling kits are addressed in detail beginning on page 128.

## Weighing the Hops

Once the bale was removed from the press and the ends sewn on, the owner used the one other piece of equipment required in the baling room, which was a scale. The owners always weighed their bales, which they did on a common grain scale. Such scales are often referred to as "hop scales" in the hop country, but they are not special tools just for hops. A 19th-century illustration shows how they were weighed, with the bale resting across the scales. (See illustration on page 112.)

The bale would be stenciled with the owner's name, post office, county and state. The weight was not stenciled on the bale. Often a grower would keep a list of his bales on the walls of the hop house. The year would be at the top, and the bale weights pressed for that year would follow.

The ideal weight for a bale was 185 pounds. By examining lists of bale weights in the old kilns it is clear that the actual weights varied, from nearly 170 pounds, to slightly more than 200 pounds. The actual weight of the hops was about 7 pounds less than the bale weight. The hop sacking and the wooden pins used in the end weighed 7 pounds. This weight was referred to as the "tare," and would be deducted by the purchaser of the hops.

## Hop Samplers

With the completion of the baling, the most critical event of the hop grower's year would begin. It was now time to sell the hop crop. Many hop growers believed that the crop should be sold by Election Day, but that did not always happen. In December of 1894, Leon Van Patten sold "old hops" for $0.035 per pound. Since old hops —last year's crop—had lost much of their value for brewing, the price was always low. Since he had 45 new bales to sell, he must have been concerned. His diary did not indicate that he sold his crop before the New Year. As with other parts of the hop grower's year, special tools were associated with the selling of the hops.

Hops were sold in two ways—either by contract before the harvest—or after the harvest, based on the market price. The best prices usually came in the months after the hops were picked. During the winter the price was usually low. The spring months often saw prices make a slight recovery. Whenever the hops were sold, a sampling of the crop was necessary to grade the hops and establish the price.

Hop bales were sampled by the use of a hop sampler. This was a tool designed to take a sample block from the side of a bale. The sampler had two handles that were hinged and extended into forks at the working end of the tool. The forks usually had three tines that were set into a bar that was perpendicular to the handles. The sampler was fairly strongly built, and came in a variety of styles.

Over the years I have seen a variety of samplers. The first time I was made aware of this tool was when I purchased a sampling kit in Windsor. That kit I sold to the State Museum in Albany. The oldest example in my collection is from Middlefield. It is a very large sampler, being 21 inches long, with tines $7^{3/4}$ inches long, and an opening of 5 inches. It has very straight handles, with wooden grips riveted to the handles. A partial mark is on one handle. It reads "William —, London." (The last name is not clear.) The only other marked sampler I have is a manufactured tool with the name "J. Murphy, 134 Broad St. NY." Other samplers have the owners initials struck into the handles, or notches on the fork that indicated the owner.

J. Murphy hop sampler.

As I said before, the largest sampler is 21 inches long. The smallest is 14 inches. The length of the three tines also varied, from 5 to $7^{3/4}$ inches, with lengths of 6 and 7 inches being most common. When opened, the distance between the forks varied as well, from 3 to 5 inches, with much variety in between. Of the ten samplers that I own, two are hand-forged, the rest are manufactured. One has a stop to limit the opening of the forks. One has a wing-nut to secure the two pieces together. In all other cases, a rivet hinges the two arms. (Several of these samplers are pictured on the following pages.)

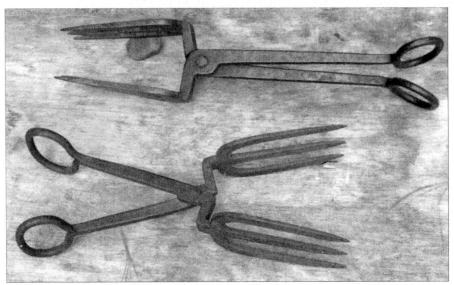
Round-handled hop samplers, Madison County. Top sampler 19 inc., bottom sampler, wrought iron, 16 in.

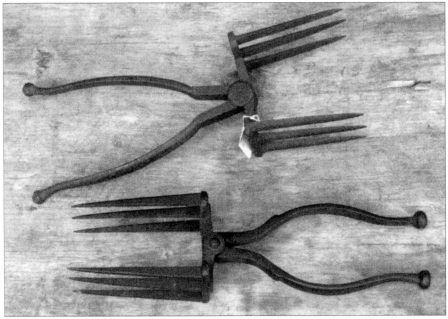
Hop samplers with ball-tipped handles, Delaware County, both 14 in.

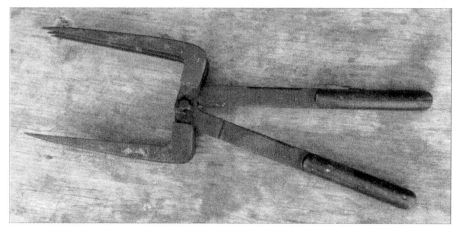
Hop sampler with wooden handles, 21 in., Middlefield, Otsego County.

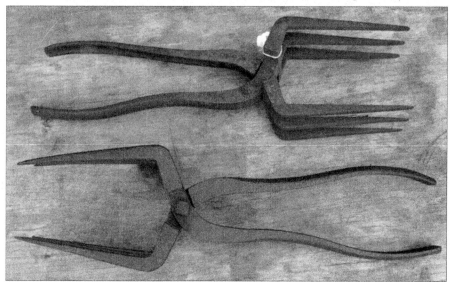
Hop samplers, both nearly 20 in. long. The bottom sampler was found in an antique shop in Weedsport, marked as a "potato cooker."

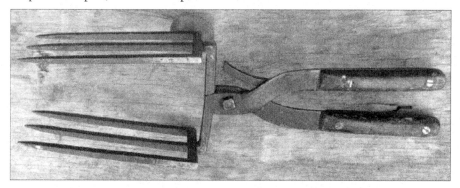
Hop sampler with unique stopping device on handle, Chenango County.

Wing-nut hop sampler, Madison, 15 in.

Two-tine hop sampler, King farm, south of Cooperstown, Otsego County. This is the only known example of this form. From the collection of David Petri.

The sampler was used to remove a block of hops from the side of a bale. During storage and until sold, the bales were placed on end, not touching one another. When a sample was taken, the bale was placed on the floor on its long side with the seamless side up. The bale cloth would then be cut open with a knife. It might be held back with bale pins, but this was not always done. Next, with the knife, a square cut was made into the bale. Now the hop sampler was taken and opened so that its tongs were placed in the knife cuts. Using one's foot, the sampler was pushed into the bale as deep as the tines. This was like the motion used in pushing a spade into the ground. With two hands, the forks were compressed and slightly pulled upward. This was done a second time on the other side of the cut. Finally, holding the sampler tight, it would be removed from the bale. The resulting block would be the sample used to determine the value of the hops.

The only known picture of a man using a hop sampler is that of the Rose family of Milford Center. The picture is a view of their hop dealership office in Portlandville. The man at center-right, leaning over the hop bale, is using a sampler to remove a sample. Notice that the boy leaning on the bale has a knife in his hands. On the bales in the back of the room are four kits to hold sampling tools. On two of the kits are two separate types of samplers. The four men on the left are inspecting hop samples and taking notes on their quality. The bale on the left is on a scale to be weighed. On the bales in the back, the owner's stencils can be seen. Robert M. Rose was a hop buyer for twenty-eight years, starting in 1876. He worked for a number of firms, starting with A.H. Akin of New York. He also worked for local firms, such as David Wilber, Charles Green & Son, and Brainard & Company. In the later years of his career, he made seven different trips to the Pacific Coast to buy for New York firms. He died on March 20, 1904, at the age of 61.

Photograph of a man sampling hops in the Rose family hop dealership, Portlandville.

The taking of samples caused some controversy among the growers. Many growers did not like to have their bales damaged by the sampling. If not done with care, some of the hops could also be destroyed. Even more controversial were the size of the sample taken, and the taking of too many samples. Growers felt that a buyer might accumulate samples to make extra bales for himself.

When I first started to collect hop tools in the 1960s, samplers were always considered to be tools used only by the buyer. Over the years this concept has been found to be in error. The buyers owned the sample kits, but they were also owned by the grower. They were sold in the hardware stores and available to all. In the 1970s I saw eight samplers in Buster Campbell's antique shop in Hyde Park, just outside of Cooperstown. He told me that they all came from the contents of an old store that he was buying. I have also talked to many farmers that remember that they had a sampler in their family, but most have lost them over the years. I have also been told that the grower often would take his own sample to a buyer to get a price. Therefore they had to have a sampling tool.

## Hop Triers

The second specialty tool used in hop sampling was the hop trier (sometimes spelled "tryer"). The trier had a long shank or handle, cylindrical in form, with a hand grip on one end, and a spear-shaped head with a barb on the other end. Two types of hand grips are found. One is a simple straight bar (or "T") perpendicular to the handle, the other is a round, open grip. The length of these tools varied, from $14^{1/2}$ inches, to 20 inches. All triers had a sharp point at the end of the head. In addition, the head had a barb that projected at about 45 degrees from the head back toward the hand grip. This barb was either squared off or rounded at its end. Some triers were hand forged, but most are factory made. I have not seen a marked trier.

The trier was used to check on the quality of the hops that a buyer had purchased. The samples made with the hop sampler would be placed on a table. Each bale in the lot purchased would be checked with the use of the trier. A cut would be made in the hop sacking about 2 inches long. The trier was thrust into the bale, given half a twist, and withdrawn. The barb would bring out a good handful of hops. These were examined and compared to the samples on the table. In this way the buyer was sure that the whole lot he had bought was of equal quality.

Many stories were told of growers adding various materials to the hops to add weight to the bale. Corn meal and dirt were two favorite culprits in these stories.

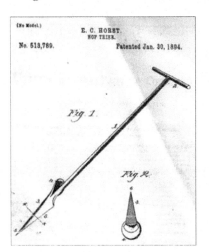

Some even tell of broken plow parts in the end of the bale to add weight. A more serious concern was that poorly dried hops might be mixed in with the quality hops. A grower would try to avoid mixing hops of diverse quality. All hops sold had to be of the quality of the sample or higher—never lower.

Emil C. Horst of Sacramento, California, patented an interesting improvement in the hop trier in 1894. He added a sharp blade to the underside of the head opposite the barb. Using this blade, the hop sacking could be cut and the trier pushed into the bale in one step. This was thought to be a time-saver.

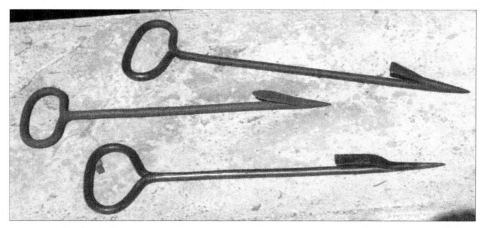

Round-handled hop triers, Otsego County. The trier in the center is hand wrought.

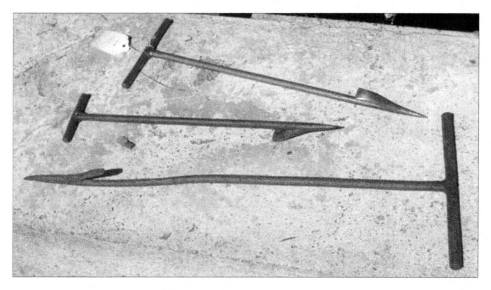

T-handled hop triers. The top trier is manufactured, the bottom two are blacksmith made. The bottom trier is from Middlefield, Otsego County.

# Hop Sampling Kits

Often a sampler and trier are found in a sampling kit, which I mentioned above. These are made of wood, leather or leather-covered wood. Usually they have a hinged top with a lock, or two hooks and eyes to fasten the lid. In the center of the top of the box is a carrying handle. Both buyers and growers used these kits. Along with the sampler and trier, other tools could be inside the kit. Often hop needles and some heavy-duty thread would be in the kit to sew the opened bale shut after the sample was removed. In some kits bale pins are also found. These were used to hold the hop sacking back while the sample was removed. Also a good, long-bladed knife would have been a handy tool. The knife was used to start the cut for the sample that the hop sampler removed. In the picture of the Rose hop office (page 125), several of these kits can be seen on the bales in the rear of the room.

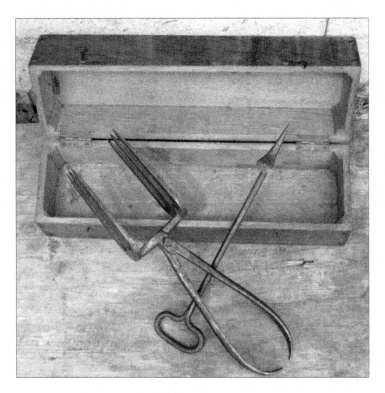

Hop sampler kit with sampler and trier, Schoharie County. The box is 19 in. long by 6 in. wide by 7 in. deep.

*Bailing & Selling* 129

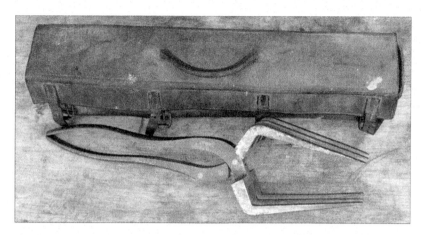

Leather carrying case for the hop sampler shown from Chenango County. Case is 21 long by 4 in. wide by 5 in. deep.

Hop sampling kit made of basswood and painted black, Cobleskill. The kit includes a sampler, a trier, a knife and several bale pins.

As the end of the year approached, the hop farmer had a few months in which he had little to do in the hop area. Only the selling in November and December would have been pressing. Other than that, the year's cycle was complete, and he could look forward to the possible riches to be made on next year's crop.

Business cards of hop merchants in the counties of Madison and Otsego.

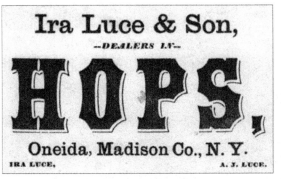

Wilber paperweight given as a gift to hop growers.

Eddy and Wilber's Hop Store, Milford.

# Appendix: Hop Houses of Central New York

One of the last remnants of the great days of hop growing in Upstate New York are the hop houses that might still be found on many farms. These buildings were a very important part of the hop culture of the 19th and early 20th centuries. Today they are an endangered species, and in some areas almost gone, but still remind us of that by-gone hop era.

The hop house, hop "kiln" or "kill," differs from most farm structures because it was a building used to cure a product—hops—and prepare them for market. Today, some try to call these buildings "hops barns," which is a totally incorrect name. Any barn is a place to store agricultural products like hay. With the exception of the off-season, hop houses were not storage buildings, but primarily processing plants.

In 1883, W.A. Lawrence of Waterville in Oneida County described the workings of a hop house as follows:

> "… every hop kiln is not only a drying house, but is also a bleachery; a preserving and curing-house, and a packing-house, all in one."

Since this building had these diverse functions, we should examine its structure before moving on to the various styles. Each hop house has two distinct sections. First, the kiln, with the furnace (or stove) room on the ground floor, and the drying room above. Second, the storage area, with the press room on the ground floor, and a storage room above.

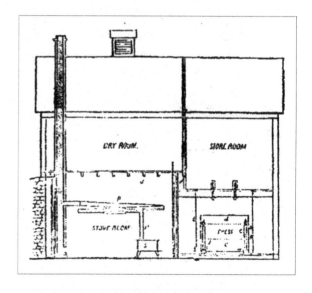

Diagram of a hop house showing the four basic parts: Furnace Room, Drying Room, Storage Room, and Press Room.

These buildings have several characteristics that make them easily identified. Each hop house has something that makes it different and unique, but all share common traits. To start with, a hop house will just have something unique about it that makes it look different than a common barn, and which will inspire further investigation. For example, the presence of a bricked-in hole for a chimney pipe would be a good sign. If you are lucky and a board of siding is missing, you might see lath and plaster on the inside of the wall. Walls covered in metal or asbestos cloth would also indicate a hop house. If you go into the building, in all styles, the presence of the basic floor plan explained above would be a dead giveaway. The presence of a slatted drying floor would also be a good sign.

Illustration of a stove room, from *Frank Leslie's Popular Monthly*.

What is pictured and discussed on the following pages is not a complete study on hop houses, but a sampling of what once could—and might still—be seen in the historic hop country of Central New York.

# Common Hop Houses

Today, the most common style of hop house found is a simple rectangular structure that is slightly tall for its size. Generally the silhouette of the building is often tall and long for the width of the building. If a cupola is present, it will be off-center and only one will be found. In a few buildings, you might see a chimney on the end or towards the center. The foundation of the building might have rectangular windows present on one end only.

Two examples of a common hop house. The one above was off Golf Course Road, Town of Cobleskill, and is no longer standing. The one below is in Engleville, Town of Sharon, Schoharie County.

Common hop house with rectangular ventilator, Cherry Valley.

Early-style hop house, c. 1860. The kiln is on the left side. This is the hop house with the metal screw press discussed on page 108.

# Step-up Kilns

A second style of hop house is the "step-up" kiln. This style gets its name from the roof line that is at two levels. Very often this indicates that the building was once in the common style, but later a new drying floor was added and the old kiln became the storage and pressing area. The taller drying area was felt to draft better than the lower common style. Many of these kilns also reflect the prosperity of the hop market in the 1880s.

Aspinwall kiln, Route 166 near Westville.

Step-up hop house, Town of Middlefield. The lower part on the left was the original hop house, the higher part was added in the 1880s.

Step-up hop house, Town of Middlefield. The kiln is covered inside with asbestos cloth.

Step-up hop house near Worcester. The kiln roof is perpendicular to the storage area roof line.

# Pyramid Kilns

The third major style is the "pyramid" style. These kilns are easy to identify, as no other type of building has the pyramid-shaped roof. Today pyramidal hop houses are few and far between, but once they were quite common in all the hop areas. Again, as with the two types above, the same characteristics exist.

Pyramidal hop house near Westford, Otsego County. The left hand side was an early hop house, with the pyramid added later.

Pyramidal hop house, Marshville, Montgomery County.

Pyramidal hop house, Middlefield Center Road, Otsego County.

These photos from the collection of Richard Vang show a pyramidal hop house that once stood about two hundred yards off the south-east shore of Canadarago Lake, near the hamlet of Schuyler Lake. This building was unique in the fact that the pyramid kiln was on the side of the storage area, rather than on the end like most pyramidals. It had an interior staircase that led to the storage room on the second floor. Notice the hole for the chimney pipe that can be seen through the door to the stove room underneath the stairs. The hole in the storage room floor for the press was still intact, and the lath and plaster of the walls can be clearly seen in the photo of the stove room. The illustration at the bottom right is the drying room of a pyramid kiln, from *Frank Leslie's Popular Monthly*.

## Round Kilns

Finally, we have the round hop houses, copied after the oast houses of the hop country in England. They are all stone and present a distinct shape on the landscape. One time I was in the hamlet of Augusta in Madison County looking at hop stove, and was told about a "hop silo" that once was in the area. It was a local term for a round hop kiln.

Cobblestone round hop house with detached storage and press room, Route 8 north of Bridgewater.

Round stone hop house, Town of Stockbridge, Madison County.

## Unique Styles

In addition to the styles above, some rather interesting hop houses exist, or did exist, that deserve special mention. Near Cherry Valley in Otsego County was a very unique double-pyramid kiln, with a huge storage area.

In the Town of Springfield in Otsego County was the Smith kiln that was made of pressed-earth bricks. This was a fairly early kiln, with a screw press inside. It was taken down in 1976 because the bricks had eroded so badly on the west side.

Several stone or partial stone kilns are also known. In Van Deusenville in Otsego County, a stone furnace room and drying floor are linked to a large wooden storage area. This kiln has one of the best screw beams and screws in the state.

An all-stone kiln is found at Cook Corners in Montgomery County. It has the date of 1862 on its interior and is probably a little older.

## Cowls

Another unique characteristic in a hop house is the presence of a cowl rather than a cupola. I have only seen this on one kiln, and it is now down.

Cowl on a step-up kiln.

Double hop house near Hubbardsville, with a step-up kiln on the left and pyramidal kiln with a cowl on the right.

# Large Hop Farms

The following pages have images of a few larger hop farms that were in the region.

View of a hop farm in Oneida County. On the left is the hop yard, with a double pyramidal kiln with cowls in the background.

Pindar's hop yard outside Middleburgh, Schoharie County. There are two double hop houses with rectangular drying floors. In the background are picked hop yards with the poles on the ground.

Jimmy Clark's "Hop City," south of Cooperstown. Today this is Route 28, looking towards Milford.

# Appendix

Onieda Chief hop house, Route 8 south of Bridgewater. At the front are two drying floors with a large storage area and press room in the middle.

Brick kiln attached to the rear of the storage and press rooms.

This is a recently-acquired photograph of Cooperstown, looking south-east from the top of the hill that is now behind The Farmers' Museum. The date of this photo is not exactly known, probably during the 1870s. While it is not an image of a single large hop farm, it shows two smaller yards at the middle left and right. It must be early in the season because the poles are set but no hops have grown yet, and there are stacks of poles next to the yard on the right. This is a good example of the small-scale operations of growers during the state's "Golden Age" of hops.

# Notes on Further Reading

Barth, Heinrich J., Christiane Klinke, and Claus Schmidt. *The Hop Atlas: The History and Geography of the Cultivated Plant.* Nuremberg, Germany: Joh. Barth, 1994.

> A good overview of the hop history in nations all around the world. The New York map is poorly done.

Bullard, Albert C. "The Hop Culture of Milford Township." In *Time Once Past Never Returns: A History of the Town of Milford, New York, 1776-1996*, compiled and edited by Linda Norris, 91-102. Milford, N.Y.: Greater Milford Historical Association, 1996.

> An in-depth study of 19th and early 20th century hop growing in Milford. Some local pictures and statistics on hop production in the township.

Bullard, Sandra Martin. *Hop Time!* Cooperstown, N.Y.: Barton-Butler Graphics, 1998.

> A brief account of hop growing based on family diaries from the Middlefield and Milford areas. It has many good pictures of tools and hop houses.

Meeker, Ezra. *Hop Culture in the United States.* Puyallup, Washington Territory: E. Meeker & Company, 1883.

> A hard book to find in the first edition, but now available in reprint. It is mostly about hop growing on the West Coast. Chapter 14, written by W.A. Lawrence of Waterville, is a very interesting article on hops in New York State.

Myrick, Herbert. *The Hop: Its Culture and Cure, Marketing and Manufacture.* New York: Orange Judd Company, 1899.

> A good discussion of how hops were grown at the turn of the 20th century. It has limited pictures and drawings of tools.

Tomlan, Michael A. *Tinged With Gold: Hop Culture in the United States.* Athens: University of Georgia Press, 1992.

> The best history of hops in the United States, with very good explanations of the plant and various sections of the country. There are some tools pictured and mentioned. There is also some good material on hop houses. Overall this is excellent, a must read.

# About the Author

ALBERT C. BULLARD earned his B.A. in History from Lebanon Valley College, and his M.A. in Folklife Studies and Museum Management from the Cooperstown Graduate Program of the State University of New York at Oneonta. From 1968 to 2001, he was a teacher at the Cooperstown Central School. His interest in hops and hop growing started while in graduate school, and since then he has collected the stories, tools and artifacts of New York's hop heritage. He has curated and contributed artifacts to various museum exhibits, most notably "When Hops Were King" at The Farmers' Museum in Cooperstown, which ran from 1998 to 2003. He has written extensively on the subject, and has given lectures and participated in seminars and hop festivals. For his activities he was awarded the title of "Hop King" at the Madison County (N.Y.) Hop Fest in 2002.

CPSIA information can be obtained at www.ICGtesting.com
Printed in the USA
BVOW09s2029041016

463745BV00026B/27/P